AUDITORY
PERCEPTION OF
SPEECH

An Introduction to Principles and Problems

DEREK A. SANDERS
State University of New York at Buffalo

PRENTICE-HALL, INC., ENGLEWOOD CLIFFS, NEW JERSEY 07632

Library of Congress Cataloging in Publication Data

SANDERS, DEREK A.
  Auditory perception of speech.

  Includes bibliographies and index.
  1. Speech perception. 2. Auditory perception.
3. Sound. I. Title.
QP464.S26   152.1′5   76–27320
ISBN 0–13–052787–4

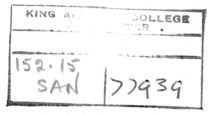

For material on the following pages; 126, 153, 154, 220, 221, 222, 224. Reprinted from, *Language by Ear and by Eye*. J. R. Kavanagh and I. G. Mattingly (eds.), by permission of The M.I.T. Press, Cambridge, Massachusetts. © 1972 by The M.I.T. Press, Cambridge, Mass.

For the material on the following pages; 91, 104, 105, 107, 125, 126, 127, 128, 129, 137. Reprinted from, *Models for the Perception of Speech and Visual Form*. W. Wathen-Dunn (ed.), by permission of The M.I.T. Press, Cambridge, Massachusetts. © 1967 by the M.I.T. Press, Cambridge, Mass.

© 1977 by Prentice-Hall, Inc., Englewood Cliffs, New Jersey 07632

Printed in the United States of America

10   9   8   7   6   5   4   3   2   1

PRENTICE-HALL INTERNATIONAL, INC., *London*
PRENTICE-HALL OF AUSTRALIA PTY. LIMITED, *Sydney*
PRENTICE-HALL OF CANADA, LTD., *Toronto*
PRENTICE-HALL OF INDIA PRIVATE LIMITED, *New Delhi*
PRENTICE-HALL OF JAPAN, INC., *Tokyo*
PRENTICE-HALL OF SOUTHEAST ASIA PTE. LTD., *Singapore*
WHITEHALL BOOKS LIMITED, *Wellington, New Zealand*

Contents

Foreword

This is a timely book. It comes at a period when professional interest is growing in relation to the issues of perception as this complicated psychophysical problem pertains to both adults and children. There have been some precursors to this volume by Derek Sanders, particularly the work of Joseph Wepman and his colleagues at the University of Chicago. Others have made contributions as the author appropriately notes.

Since 1945, interest in perception and perceptual processing has become much more widespread than earlier. With adults the interest in part was initiated by the large number of men returning from the Second World War with neurological injuries. With adults this interest and professional concern has continued through successive wars in Korea and Vietnam. The knowledge and research accrued from these sources has, of course, spread to the adult population generally.

With children, there has been wide-spread development generally conceptualized under the term, learning disabilities. More appropriately these are children with learning disabilities which are the outgrowth of perceptual processing disabilities which in turn are for the most part of a neurophysiological origin. These children are variously known and characterized in the literature. Regardless of the terms used to describe them, their problems are essentially those related to perceptual processing.

Since 1937, although some would indicate before that date, the em-

phasis on this matter has been primarily related to visual perceptual processing problems in children of several clinical categories, i.e., exogenous
mentally retarded children; children with cerebral palsy, epilepsy, encepha-
litis; children characterized by hyperactivity and failure to control attention; children with aphasia; and to a lesser extent with children who have
tactual preceptual processing problems in conjunction with blindness. With
the exception of the latter group and the studies of Mildred McGinnis
with aphasic children, the research related to the above-mentioned categories has essentially been of visual-perceptual-processing nature. Other
sensory modalities have received little attention from researchers.

The research in the field of visual perceptual processing deficits is
minimal; that in the other sensory modalities, even less. In areas other than
vision, this may be due to the difficulty and expense of developing techniques and equipment with which to perform adequate research. It is also
due undoubtedly to the small number of professionally prepared research
personnel in the field who are able to undertake knowledgeably that which
must be done. Audiologists, the logical group to undertake auditory perceptual processing research, have for the most part been more interested in the
measurement of hearing acuity than in auditory differential diagnosis in
children or adults where auditory acuity per se may not be the essential
problem. Exceedingly few studies are available which concern themselves
with perceptual processing problems in tactual (haptic), gustatory or olfac-
tory modalities. The significant exploratory work of James L. Paul concerned with the responsivity of brain-injured children to changes in relative
humidity and temperature has not been replicated a decade later, although
this information is important in planning for the growth and development
of these children.

There is a vast area of human development, longitudinal and cross
sectional at varying chronological ages, which must be done in order
eventually to fully understand the adjustive mechanisms of a large and
undetermined number of children and adults.

This volume begins to fill the vacuum which exists in the area of
auditory perception. It is a focus both on the physiological aspects of
perception as well as the neuropsychological aspects of this sensory
modality,—an aspect of development obviously significant in the adjustment process of the human. Developed within a theoretical position, the
contents of this volume constitute an important link to that which has
been done in the field of auditory perception as well as pointing directions
to those things which yet must be done.

William James many years ago stated that "The intellectual life of
man consists almost wholly in his substitution of conceptual order for the
perceptual order in which his experience originally comes." While the
present generation may have different terms and while James did not have

the perceptually handicapped individual in mind when he wrote this statement, nevertheless his observation is essentially true. The role of auditory perception in both perceptual and conceptual learning (particularly in the acquisition and use of langauge) has been recognized for many years. Its significance is especially noted when maldevelopment takes place as in the child with a physiological problem resulting in a measured hearing loss or in a psychoneurological disturbance leading to an auditory perceptual processing problem characteristic of many children with learning disabilities. Aphasia is undoubtedly the extreme form of this.

To state that research is needed is to state the obvious. The most significant issue pertains, not to theoretical structure of research, but to the methodology. It is relatively easy, in spite of the complications of equipment and technology, to perform research on auditory perceptual processing, Such research is essentially of an external nature, i.e., external to the human organism and its neural system. The basic problems remain unsolved by such external "measurement" research. There are no ways available as yet for research personnel to study with assurance that which goes on within the lobes of the human brain itself during conditions of life, within either the central or pyramidal aspect of the nervous system. Although there are educated guesses, that which goes on at the synapse under conditions of learning (even in the most primative forms of learning) is essentially unknown. A study of the human nervous system under laboratory conditions adds much to the knowledge of its operation, but there is a significant difference between the study of the brain under these circumstances and the study of brain under conditions of life when electrical energy is surging in its complex ways throughout the brain structure. The inability to study the mechanism basic to all perceptual processing will produce problems for researchers—neurologists, psychologists, neurophysiologists, and others—for a number of years to come.

This volume adds to the literature in this field and assists those concerned with developmental psychology and developmental neurology to understand their task better. More books and professional articles based on research and good theory are needed in the field of auditory perception as well as in the other modality areas which we have mentioned. This volume and others like it will permit practitioners who assist children and adults to have a logical theoretical orientation and stance from which to work.

WILLIAM M. CRUICKSHANK

University of Michigan

Acknowledgments

The completion of this book has been possible because of the support and assistance given so generously by many people. I wish to thank all who have shared in its preparation. Special thanks are due to Robert E. McGlone for providing many original illustrations for Chapter 2, and to Berner Chesnutt who gave me valuable advice in preparing the text of Chapter 3. The Educational Communications Center of my university made available the skilled services of Barbara E. Evans, who drew the many line drawings throughout the book, Melford Diedrick, medical illustrator, who did the artwork for Chapter 3, and Jim Ulrich who provided extensive photographic work.

Many of the changes I made in the final draft arose from the insightful guidance given by Fred Gruber, and Dr. William Cruickshank, to whom I also owe my appreciation for the foreword.

I wish to say a very special thank you to my graduate assistants Jan Silverman and Marsha Shapiro who worked so hard on the typing and critiquing of the early chapters and to Judy Hirsch, without whose moral support and extensive help in typing and editing I would not have met publication deadlines. In the tedious task of editing I received much help from Judy and from Janet Genchur.

Finally, and most importantly, I thank my wife Cynthia and my daughters Jennifer and Hilary for their understanding during the difficult periods which accompany the writing of any book.

DEREK A. SANDERS

Buffalo, New York

A theory is a theory, not a reality. All a
theory can do is remind me of certain
thoughts that were a part of my reality
then. A statement or a "fact" is an
emphasis—one way of looking at something.
At worst it is a kind of myopia. A name
is also just one way of seeing something.

I can't make a statement about a reality
without omitting many other things which
are also true about it. Even if it were
possible to say everything that is true about
a reality, I still would not have the reality;
I would only have the words. In fact, the
reality changes even as I talk about it.

When I outgrow my names and facts and
theories, or when reality leaves them behind,
I become dead if I don't go on to new
ways of seeing things.

Hugh Prather, *Notes to Myself*. (Moab, Utah: Real People Press, 1970)

1

A General Overview

The central topic of this book is auditory perception—in particular, the perception of spoken language. We must keep in mind, however, that this particular sensory modality does not exist in isolation. Most aspects of perception are part of an integrated system of perceptual processing that has an underlying mode of function common to all. Church has stated:

> If we are to take a biological view of behavior, we must stop thinking of the organism as a collection of physiological subsystems to be studied individually in their basal or resting states and collectively only as they work together homeostatically. Instead we must see how the body's subsystems are subordinated to and organized in the service of the organism's communication with reality so that the characteristics of their functioning change in different contexts. (1961, p. 29)

The approach we shall take to the process of auditory perception is based upon the phenomenological view that knowledge of our environment derives from a reflection of what is directly available to consciousness. Our mind must deal not with real things, but with the appearances or *phenomena* of these real things, subjectively apprehended through the medium of our sensory systems. Sensori-perceptual systems, therefore,

1

play a primary role in determining not only how man adapts, but also how he perceives the stimulus to which he must adapt.

Such an approach eliminates the question of whether or not our perceptions are dependent upon an absolute reality. It also enables us to understand how our auditory system can, and does, perceive stimuli differently on different occasions—or even more puzzling, how we can perceive different stimuli as being the same, or even perceive stimuli to be present when they are not.

I shall attempt to show that we perceive the auditory world not as it is but according to how it is processed.

Let us consider the sensori-perceptual system from this point of view: We know that this network actually detects and responds to changes that occur in the physical organization of the organism's environment. When such changes take place within the sensitivity range of the sensori-perceptual systems (i.e., when stimuli impinge upon the organism), they in turn are modified in their organizational structure. It is these modifications that are apprehended by the lower neural centers and the cortex. Thus, we perceive not the world around us, but the changes it has induced in us.

Each individual perceives the external stimulus in his own terms. Furthermore, it is a fact that the sensori-perceptual system is itself capable of constantly modifying its state; thus, the perception of the external environment changes as our relationship to it changes. Does this lead us to conclude that we are not passive recipients of a flow of sensory information, but active participants in a dynamic relationship with the physical world?

I believe that we must come to this conclusion. We are bound by the facts to view the sensori-perceptual systems as active systems which select certain stimulus complexes from the ongoing array of potentially available stimuli and process them according to some previously determined criteria for relevance. Acceptance of the concept that the subsystems are organized to facilitate the organism's adaptation to reality leads us to assume that some common denominator must exist for all sensory systems. It would be most inefficient to have five sensory systems all functioning quite independently despite their common purpose. Furthermore, incompatibility in manner of function would exclude the possibility of intersensory interaction which is known to take place (see Chapter 3). The common denominator we propose in this text is that of information patterning. Although the sensory organs are structurally different to permit reception of different types of stimuli, they all share a single function: Each sensory organ has evolved to serve the purpose of internalizing data about what is happening in the environment.

Our sensory end-organs constitute an interface . . . between the subjectivity of the organism and the objectivity of the physical world. In order for any part of the physical world to be assimilated by the organism, the physical reality must first be altered to be compatible with the ability of the system to process it. Foods must be broken down through a process of digestion before they can be used as nutrition for the body, just as the air we breathe must be broken down and the oxygen changed to oxyhemoglobin before it can be utilized by the cellular system. Similarly, the physical events of the external world which impinge upon the sensori-perceptual system are themselves not capable of entering it unaltered. Yet, we know that we are able to perceive objects, persons, and events occurring around us. Clearly, something makes this possible. In order to understand how this is achieved, we must examine what occurs at the interface between the environmental system and the sensori-perceptual system.

We have seen that the sensory end-organs register changes in their respective media. This fact establishes a direct relationship between the pattern of events in the environment and the resultant pattern of events in the sensori-perceptual system. Any adaptation made within the system reflects approximately the pattern of events occurring outside the organism. Therefore, the form our perceptions take is to some degree predictable.

However, before we can make predictions we must have a certain amount of information from the object or event. The term *information*, as we shall be using it, derives from a statistical theory having to do with a measurement of probability.* Thus, information refers to any clue, or part of a clue, which constrains choice and consequently increases the probability of a correct prediction. Information is generated by those aspects of the stimulus or stimulus complex which serve to constrain or limit the predictions made by the sensori-perceptual system. These complexes arise from objects or events which are themselves organized, or *patterned*. By virtue of their structure, the stimulus sources pattern the environmental changes and thus limit the response behaviors of the sensory end-organs. Gibson stresses the importance of making a clear distinction between the *sources* of stimulation and the *stimuli* themselves.

> The former are objects, events, surfaces, places, substances, pictures, and other animals. The latter are patterns and transformations of energy at receptors. A stimulus may specify its source, but it is clearly not the same thing as its source. (1966, p. 28)

It is these patterns and transformations which generate information. At the basis of all perception rests the ability to identify or impose patterning upon the changing sensory environment. Our concern will be with the

---

*For a review of this theory see C. F. Hockett, "The Mathematical Theory of Communication," in *Psycholinguistics,* (ed.) Sol Saporta (New York: Holt, Rinehart and Winston, 1961), pp. 44–67.

particular patterning present in the speech stimulus and the patterns we as listeners are able to impose upon the incoming speech stimuli.

Thus, we can say that patterning generates information. It serves to constrain the organism in the perceptions it will experience as a result of changes induced upon it by the interaction of the organism with its environment. These constraints arise from the interaction of factors from within the organism as well as from the environment.

Gibson (1966) has pointed out that all things broadcast information about themselves when they are in a medium capable of conducting the flow of information. Just as the information flow from the medium to the organism is dependent upon the receptivity, or sensitivity, of the mediating sensori-perceptual organs, so is the source dependent upon a receptive medium for the transmission of information. A ringing bell will be seen but not heard in a vacuum because the medium is not responsive to the vibrations of the bell. Providing the medium is appropriate, information always will be potentially available to organisms sensitive to it. As Gibson has noted:

> Most of what is broadcast by a light-emitting or light-reflecting source, or a noisy, or an odorous source, is never picked up at all; it is wasted for purposes of stimulation. But a set of perspective projections, or a field of a volatile substance is a perfectly objective physical fact. It is *potentially* stimulating, and whether or not it actually excites receptors need not here concern us. (1966, p. 14)

Certain limitations exist, or may occur, at various stages of the process of perception and communication to limit the range of potential stimuli. In general these limitations fall into one of two categories: (1) the receptive system or (2) the relevance of the stimulus array. Auditory perception results from the interaction of these two factors in a selective manner under the control of the perceptual system.

In order to develop an understanding of how this process occurs, it is desirable to develop a concept that will serve to integrate the various components of the auditory pattern. To be valid this concept must be applicable to all aspects of the process. It must facilitate our understanding of the acoustic signal of speech as much as it does that of the syntactic processing of the speech signal. The concept we shall adopt to perform this task is *pattern processing*.

A pattern is not in itself a physical entity but an expression of the relationships between a number of component entities. As such it changes as those relationships change; conversely, providing the relationships remain constant, the pattern remains constant, despite changes in the actual physical components. This fact is important when we explain perceptual invariance in the presence of a variable signal. Most important to our theory is the fact that a pattern is not subject to transmission difficulties between different media.

It will be our aim, therefore, to examine each stage of the process of speech communication to determine the influence of patterning. We will begin by examining the physical entity of the acoustic signal.

The Acoustic Signal

Sound waves are patterns. They are measurable, physical entities which, by their patterning, permit potential information to travel between the sound sources and the receiver. Most texts dealing with the physical parameters of sound devote a great deal of attention to a discussion of sound transmission, pure tones, simple and harmonic motion, overtones, and complex wave analysis. Our interest will be less with the physics of sound and more with the *relationships* between the physical characteristics of the acoustic wave and the individual's perception.

The speech organs pattern the movement of air molecules (speech sounds) according to rules which link the language system to the speech production system. The process of turning language into speech is a continuous one. It involves the generation of changing patterns of complex pressure waves. These wave patterns are the constraints which provide the listener with a key to the linguistic rules used by the speaker to generate the code. Once the key has been correctly identified, it can be used to break the code in order to reconstruct the intended meaning.

Of course, speech is not the only way man constrains the environment to transmit information. He does so when he writes, acts, makes music, dances, draws, paints, or arranges flowers. Edward Hill, for example, writes about the language of drawing:

> Drawing diagrams experience. It is a transposition and a solidification of the mind's perceptions. From this we see drawing not simply as a gesture, but as mediator, as a visual thought process which enables the artist to transform into an ordered consequence what he perceives in common (or visionary) experience. For the artist, drawing is actually a form of experiencing, a way of measuring the proportions of existence at a particular moment. (1966, p. 8)

When we look at a drawing, we are subject to the constraints which the artist has placed on paper. Reaction to the external, physical constraints of the light-reflecting properties of lines and shading will be different for each of us. Some people will require more information from the source than others in order to perceive some aspect of the picture; some will be more sensitive than others to the subtleties of information content; some will contribute a great deal of information to the experience; others will be limited in their perceptions. Nevertheless, to the extent that there is compatibility between the organism and the potential information generated by the source, information will have flowed from the artist to the observer.

Speech too serves as a mediator. Hill suggests that the artist relies upon his ability to reproduce in graphic form information he believes in some ways to be equivalent to what he is seeing. Similarly, some researchers suggest that the process of auditory perception involves the mediative use of, or reference to, the particular articulatory movements which have been used to imprint constraints on the acoustic signal.

Although it does not appear that any one particular theory is yet capable of accounting for speech perception, the motor theory (Chapter 5, pp. 116–124), which emphasizes the role of the production stage of speech communication, has made significant inroads into present thinking on auditory perception and has provided a new way of viewing the acoustic structure of speech. It is the patterned information in the acoustic signal that the auditory system must internalize. Our task now is to determine the operation of this system.

The Auditory System

Just as the discussion of the acoustic signal can be approached from the standpoint of its relationship to pattern perception, so can the anatomy and physiology of the auditory system. Instead of examining the hearing mechanism as a structure, we will view both the peripheral and central systems in terms of the role they play in detecting and analyzing the incoming patterns of acoustic information. The concept of the perceptual systems as active processors of information patterns is supported by Gibson. He maintains that we must recognize them as

. . . neither passive senses, nor channels of sensory quality, but ways of paying attention to whatever is constant in the changing stimulation. (1966, p. 4)

The concept of an active auditory system directly influences the way we view the processes of speech perception. It necessitates not only that we be familiar with the anatomical and neurological components of the auditory system, but also that we understand how they function as pattern processors. This information is critical to an understanding of neurologically based theories of speech perception.

The Perceptual Process

When we view perception as the function of an integrated system, we necessarily assume that certain fundamental processes are basic to all sensori-perceptual modalities. For example, any theory that attempts to explain how patterned information is extracted from an incoming visual signal can be applied just as well to auditory processing.

One of the measures of strength of a theory rests in the breadth of its applicability. For this reason, our pursuit of an understanding of auditory perception must lead us first to examine theories dealing with the sensory processing of information. This knowledge should pave the way for our study of those theories developed specifically to explain auditory perception of speech.

Theories of Speech Perception

Theories of speech perception can be divided into two classes: *passive theories*, which conceive of a direct relationship between the acoustic pattern and neural representation of that pattern; and *active theories*, which propose an intermediary process to recode the information in the acoustic event prior to its perception as a spoken language code. Rather than elect complete support for either one of these two positions, we shall examine the writings of the exponents of each in an attempt to find evidence for a view that accommodates both of these apparently conflicting theories.

Language Constraints on Auditory Processing

The perception of speech in all but the experimental situation involves the perception of spoken language. One might say that speech is in fact a symptom of language and is thus inextricably a part of it. The nature of that association is crucial to our understanding of how we perceive speech. It will be particularly relevant in considering auditory learning difficulties involving speech production problems, reading difficulties, and receptive problems in processing spoken information.

The first question which the speech/language relationship raises is how we are able to identify discrete linguistic units from the continuous stream of information reaching our ears. This process requires segmentation of the acoustic stream by the listener. It is one which is heavily influenced by the use of the grammatical rules of the language by speaker and listener and by the use of the melodic components of speech. These melodic components are called *prosodic* or *suprasegmental* and are distinct from the segmental structures grammar generates.

Because both the segmental and suprasegmental information is patterned, the units of the pattern will have an order peculiar to that pattern. This plays an important role in pattern recognition. Since each component is related to those preceding and succeeding it, prediction of probable future units becomes possible. Prediction greatly enhances the rate of perception and consequently the rate at which spoken language can be perceived. The ability of the system to identify rather large segments of the pattern is essential, since it is related to the problem of storage. Storage is necessary because the speech signal is both brief and transitory. At no time do we have more than a small section of the acoustic signal before us. It is essential that the auditory system be able to hold sections of the pattern in storage to permit them to be processed into meaningful units. This involves two types of memory, short and long term, each of which involves a different storage process. Short-term memory handles unsegmented information which is then processed into long-term memory by restructuring the information into linguistic units. We shall examine the factors which influence the varying size of these units.

Once memory for language patterns has been established, the process of listening to speech is greatly simplified. The ability to predict the evolving pattern by sampling the information received and by generating probabilities concerning its ultimate form greatly reduces dependency upon the incoming signal. This makes it possible to listen to speech against a background of other noise. Many children exhibit difficulty in performing this everyday task of paying attention. It may well prove that for many the root of the problem may lie in the processes of auditory and/or visual language coding.

Auditory Perceptual Development in Infancy

Until very recently our knowledge of auditory perception has been derived exclusively from studies of perceptual behavior and from introspective reporting of adults. In the last few years there has been considerable interest in the development of research models, or *paradigms*, applicable to controlled study of the development of auditory processing of speech. The knowledge obtained to date on this topic will be summarized and discussed in Chapter 7 by Philip Morse, who has made a significant contribution to this research. The study of infant auditory development promises to contribute a great deal to our understanding of the fundamental processes that underlie speech perception. One major question under current research is to what extent the ability to perceive speech is a genetically endowed human characteristic. We are beginning to realize that at least in terms of pre-linguistic or *primary auditory* processing at the level of the acoustic signal, babies show unexpected sophistication in detecting differences between speech sounds. How these abilities are related to the acquisition of linguistic knowledge remains to be demonstrated.

It is also possible that the sensory processing capabilities of infants may provide an index of the maturity of the central nervous system (Kessen, Haith, and Salapatek, 1970). Moveover, Eisenberg (1974) and Lewis (1975) have suggested that physiological measurement of infants' auditory responses to differential stimuli may provide useful diagnostic information. Many children experience developmental problems in processing auditory language and later in learning to associate the spoken and written forms of the word. An understanding of auditory perception of speech and language during early childhood may help define the problems of auditory learning.

Defining Auditory Processing Problems

In recent years there has been growing concern for the group of children who, despite normal sensory functions and normal intelligence, experience considerable difficulty in learning situations. Among the problems these children exhibit are those associated with auditory learning tasks. Our model of auditory perception may shed some light on these auditory perceptual difficulties. Although we shall not discuss this topic extensively, we hope to indicate how major concepts of our model can be applied to these problems. It should be reassuring to approach this task equipped with current theory and research evidence on speech processing.

The value of such insight becomes immediately apparent as one confronts the literature dealing with auditory learning problems. The absence of a well-reasoned and well-substantiated theoretical model is obvious from the confusion of terms used in the classification of children having perceptual processing difficulties. However exasperating the arguments over nomenclature may be, we must not underestimate the role semantics plays in shaping our concept of a problem. Quite obviously, then, in order to proceed in the field one must become familiar with both the terminology and the various means of identifying certain symptoms in children. It is necessary, for example, to understand the differences between peripheral problems associated with hearing impairment and those involving the central processing functions. Since there is much overlap in the resultant behavioral patterns, differential diagnosis is of paramount importance. In Chapters 8 and 9 we will review the current ways of defining deviant auditory perceptual behavior and examine the compatibility of such categories with our conceptual model.

Auditory Learning Difficulties as a Language Processing Dysfunction

The auditory perceptual model has traditionally stressed the role of language in speech perception. This naturally leads us to assume that auditory learning difficulties probably originate from language coding problems. The most prevalent view in the literature on auditory perceptual disorders is that they arise from a dysfunction of a specific auditory skill or set of skills. Such a conclusion seems unjustified in the light of the theoretical and research evidence on normal and deviant processes of speech perception. Certainly, there are identifiable components or aspects of auditory processing. However, it seems more reasonable to assume that these dysfunctions are *symptoms* of underlying processes rather than the actual processes themselves. We shall examine what seem to be the major aspects of auditory processing, but it is hard to accept the existence of a set of discrete auditory skills, since each aspect constitutes only one facet of the process and intimately involves one or more other aspects. What implications performance on isolated tasks may have for an understanding of deviant function of an integrated system is something we will need to consider, for it applies not merely to speech perception and production but also to reading.

Auditory Perception and Reading

The arguments concerning whether or not auditory perceptual problems arise from deficiencies in specific auditory skills has its corollary in the area of reading difficulties. There is little doubt that auditory perception plays an important role in learning to read. Certainly, the widely used phonic approach to reading depends heavily on the child's ability to process patterns of speech sounds and their linguistic values. Examination of existing theories about how a child learns to read reveals a surprising parallel to theories we will discuss concerning how the spoken word is perceived. Many of the same questions are asked and many similar hypotheses are made concerning how the process might operate. It seems quite logical to assume that a common basis exists for an understanding both of speech perception and reading.

Intervention

All our discussions lead to the conclusion that serious doubts should exist about the validity of both diagnostic tests and remedial approaches to the management of auditory learning disorders based upon the concept of specific auditory skills. At this time, new concepts concerning auditory and visual language processing are just beginning to emerge. It is not possible at this time to provide a clearly defined and delineated alternative intervention model. The educational and clinical evidence is only just beginning to accumulate and still remains to be assessed. However, in our final chapter we shall endeavor to provide our readers with some guidelines to use in their own thinking about the origin of auditory learning difficulties.

REFERENCES

CHURCH, J., 1961. *Language and the Discovery of Reality*. New York: Random House, Inc.

EISENBERG, R. B., 1974. "Use of Electrophysiologic Measures in the Assessment of Auditory Functions," in *Sensory Capabilities of Hearing Impaired Children*, ed. R. E. Stark. Baltimore: University Park Press, pp. 23–38.

GIBSON, J. J., 1966. *The Senses Considered as Perceptual Systems*. Boston: Houghton-Mifflin.

HILL, E., 1966. *The Language of Drawing*. ("A Spectrum Book") Englewood Cliffs, N.J.: Prentice-Hall.

KESSEN, W., M. M. HARTH, P. H. SALAPATEK, 1970. "Human Infancy: A Bibliography and Guide," in *Manual of Child Psychology*, 3rd Edition, eds. L. Carmichael and P. H. Mussen. New York: Wiley.

LEWIS, M. 1975. "The Development of Attention and Perception in the Infant and Young Child," in *Perceptual and Learning Disabilities in Children*, Vol. 2, Research and Theory, eds. Cruickshank, W. M. and Hallahan, D. P. Syracuse, N.Y.: Syracuse University Press, pp. 137–162.

2

The Acoustic Signal

In the last chapter we commented that speech does not occur in a vacuum—that is, communication is part and parcel of human interaction, and only within such a context can the nature of the act of communication be understood. This statement is metaphorically true; however, it is also physically true. A vacuum deprives the sound vibrator of the medium necessary for the transmission of its energy. Sound vibrations travel with varying degrees of ease through solids, liquids, and gases. However, as far as speech communication is concerned, air constitutes the medium of sound transmission.

Sound vibrations constitute a physical event: they comprise waves of energy, generated by a vibrating body, traveling in all directions from one air molecule to the next. These energy waves move away from the vibrating object in concentric circles, becoming weaker and weaker as they move farther and farther from the source, until the energy is completely converted into another form, usually heat. The concept may be illustrated by the analogous situation in which the water molecules of a pond are repeatedly disturbed. Energy generated by stones thrown into the center of the pond moves across its surface in all directions in the form of small concentric waves increasing in diameter and decreasing in height as they move farther and farther from the center. A cork floats, rising and falling with each wave, yet remaining in essentially the same position, demonstrating that it is the energy wave and not the water which moves outward from the point of impact of the stone. Similarly, the air molecules themselves move very little, since each is restricted in its excursions by the molecular forces of adhesion. However, within the limits of their own elasticity, the molecules are capable of being moved closer together, or of being pulled apart from each other. When energy crowds air molecules together, there is an increase of air pressure around the point of greatest crowding; when the distance between the molecules is increased the air pressure decreases. Thus, sound waves are pressure waves of positive and negative value; that is to say, *pressures greater or less than the natural air pressure* (Figure 2.1).

The fact that a molecule can be moved varying amounts in a positive (crowding) or negative (scattering) direction provides the basis for the patterning of the pressure waves that constitute the information-bearing content of the sound wave. "Crowding," resulting from positive pressure, is termed *compression* while "scattering," resulting from negative pressure, is known as *rarefaction*. The role of the medium is to behave in a manner closely equivalent to the behavior of the vibrating body which is broadcasting energy waves. Thus, the pattern of movement of the vibrator is imprinted upon the medium in the form of sound waves carrying the po-

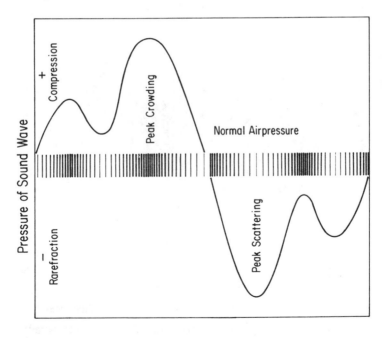

Figure 2.1 A complex waveform indicating periods of compression and rarefaction over a period of one second.

tential stimulus information away from the source. This is an important concept for, as we shall discuss later, perception is based primarily upon pattern recognition. Since we do not physically come into contact with the vibrator, faithful reproduction of its vibratory pattern within the intervening media is very important to perception (Figure 2.2).

The Perceived Characteristics of Sound

Before we consider the acoustic structure of speech sounds, we might do well to ask ourselves, what are the characteristics of the sounds we perceive? Regardless of whether we consider environmental sounds or the sounds of speech, we are able to identify certain characteristics perceptually. These include the *loudness*, from very loud to soft, the *pitch*, with variations from high to low, the *quality* of the sound which is defined by such adjectives as "harsh," "rich," "tinny," "bright," and "crisp," and finally, the length or *duration* of the sound and the manner in which it

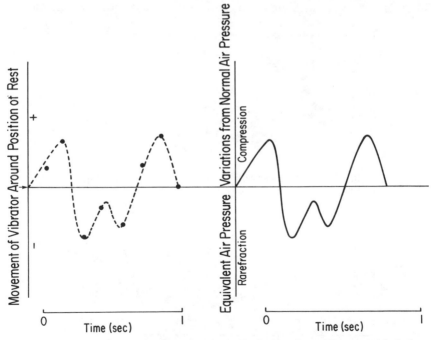

Pattern of Vibrator Movement translated into Equivalent Sound Pressure Variations

Figure 2.2 The imprinting of a vibratory pattern upon an air medium.

changes. The manner of change is quite obvious when we compare the sound of a vacuum cleaner or automobile engine with the sound of a clock ticking or a telephone ringing. Although this characteristic of change is sometimes difficult to hear, it is present in the acoustic identity of individual speech sounds and in sentences, where rhythm and stress generate additional constraining information. It is continual change which generates identifiable *acoustic patterns*, i.e., *manifestations of the interrelationships between the individual components*. The manner of change which occurs from one speech sound to the next constitutes a very important source of information referred to as a *transition*, a concept we shall examine in more detail later in this chapter.

Each of these perceived characteristics must to some degree be related to the acoustic waveform. The waveform serves to constrain the auditory system in its processing of the acoustic signal. The listener will perceive, therefore, in terms of his own system's organization rather than in terms of some received package inherent within the acoustic signal. Nevertheless, the acoustic waveform surely provides both the stimulation to auditory perception and a considerable amount of the information constraining us to experience an appropriate perception. For this reason we are justified in beginning our attempts to understand auditory perception with a selective look at the nature of the acoustic event.

The Physical Characteristics of Sound

Most vibrating objects in our environment, including the human vocal folds, vibrate in a complex manner. Complex vibrations occur because most sound sources do not vibrate as a whole (which would result in a simple pattern) but in segments. Each segment vibrates at a different rate, or *frequency*, and therefore contributes energy of different frequencies.

An air molecule is subject to several energy forces acting upon it simultaneously. Since it cannot move in several directions at the same time, it moves in a manner which is the resultant of the interaction of the multiple forces. Complex sounds thus contain a mixture of many frequencies, each of a different energy strength (Figure 2.3). The actual ingredients which produced a particular complex waveform, or pattern, can subsequently be deduced through frequency analysis conducted either mathematically (Fourier analysis) or, as is usually the case, instrumentally.

All speech sounds are complex. To a great extent speech perception involves analyzing the complexity to establish the particular arrangement of the components of the sound wave. Since the various component frequencies can be combined in an enormous number of ways, it has been

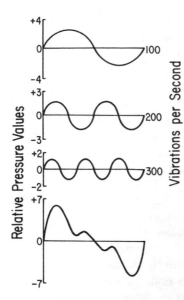

Figure 2.3   A complex wave resulting from the interaction of three simple components. (Sanders, 1971, p. 42)

possible to evolve a system of communication in which fairly distinctive acoustic patterns have been established for each phoneme of a language*

Intensity and Sound Pressure

Speech sounds are defined in terms of the three measures (parameters) of intensity, frequency, and duration, the interaction of which comprises the spectrum of the complex sound.

The first characteristic to be considered is that of intensity. The energy in a sound may be measured in terms of its intensity or its sound pressure. Both refer to the same parameter of the sound wave, but each is based upon a different reference level. Sound pressure can be measured in terms of the total amount of pressure, or it can be measured more discretely as the energy of each frequency component. The ear performs both of these analyses, providing the first level of basic information concerning the identity of the speech sound. The pressure of the speech sound is dependent upon the amount of energy applied to the vocal folds of the larynx. When a small amount of energy is applied, the extent of their movements

---

*A phoneme refers to any speech sound that in at least one instance in the spoken vocabulary of a language can be shown to differentiate one word from another. For example, in *sit-set-sat* the vowel in each case changes the meaning: each can therefore be designated a phoneme.

will be limited; when a large amount is involved, the folds will move with greater vigor. Since the vocal folds are surrounded by air molecules, their movements will disturb the air around them, and the extent of the disturbance will reflect the magnitude of the movement of the folds. In this manner the energy is projected into the medium of the air and potentially will be available to anyone within hearing range. Intensity changes of certain magnitudes produce perceived changes in loudness. Although the perception of loudness is not totally dependent upon the intensity of the sound, it is very closely related, and we can state that *loudness is the psychological correlate, or counterpart, of intensity.*

If we analyze a sample of normal conversational speech, we will find that the average pressure level of the voice at a distance of four to six feet is 60–65 dB Sound Pressure Level (SPL). The initials dB abbreviate the term *decibel*, the most common unit used to express acoustic pressure. The decibel is one-tenth of a Bel, named after Alexander Graham Bell, the inventor of both the hearing aid and the telephone. *The decibel is not an absolute measure, it is a reference scale which expresses one quantity in terms of another.* You could express in decibels, if you chose, the number of eggs laid by a particular chicken. To do so you would have to select a representative hen whose egg production figure would then serve as the reference. The output of your particular hen would then be expressed as a ratio of the output of the reference hen. An unlikely exercise, fabricated merely to dispel the notion that a decibel is a particular amount of sound energy.

In expressing sound energy we use a logarithmic computation with a unit of 10 as its base. Thus, decibels increase as multiples of the base 10 (10, 100, 1000, 10,000, etc.), rather than linearly (10, 20, 30, 40, etc.). A 20 dB sound is not twice as powerful as a 10 dB sound but one hundred times more powerful, that is, 10^2 or (10×10). An 80 dB sound is not eight times more powerful; the 8 tells us that the original power 10 dB has been multiplied by itself eight times (10^8) and is therefore one hundred million times more powerful. It is apparent from these figures that we are dealing with enormous energy ratios even though the absolute energy changes are quite small.

When we express sound levels as *acoustic pressures* we use the same decibel scale, but the method of calculation is slightly different since acoustic power and acoustic pressures are not identical. Fortunately the conversion is a simple one, for *the acoustic pressure is proportional to the square root of the corresponding power.* Thus, for acoustic pressures, a tenfold increase is 20 dB, etc.

For most of you, these physical computations are not terribly important; you should not be unduly disturbed if you have trouble with them. For our purposes it is not essential that you master the actual details of sound measurement, but only that you have a working knowledge of the terms and concepts we need in our discussion of auditory perception. Those of you who are dissatisfied with this superficial treatment of sound intensity should consult Zemlin (1968), Ladefoged (1962), and Minifie, Hixon, and Williams (1973) for a more thorough explanation.

A decibel has no fixed value, but expresses the proportional value of one sound pressure to another. Consequently, the number of decibels quoted is quite meaningless without a reference figure to serve as the acoustic pressure against which the proportional values will be fixed. The figure usually used as a reference pressure is 0.0002 microbar (or dynes/cm^2),* which approximates an average of the least amount of intensity detectable by healthy young ears under ideal listening conditions. Acoustic pressures are commonly expressed with reference to this base intensity and are identified as *Sound Pressure Levels* or *SPL*.

Let us return to our discussion of the decibel. We stated that the average sound pressure level of conversational speech at four to six feet is 60–65 dB SPL. If we analyze that energy over time, we find that it is sometimes higher and sometimes lower. These peaks and dips are evidence of the differing pressures of various speech sounds. Although it is not possible to discriminate speech sounds on the basis of sound pressures alone, it is possible to divide them into several different pressure categories.

We shall repeatedly encounter evidence that the process of speech perception is probably based upon the progressive categorization of information. Each successive analysis of the acoustic signal further rules out certain probabilities and narrows the choice of potential sounds. Differences in the sound pressures of speech sounds probably account for the first sifting out in this process of elimination. Table 2.1 (p. 19) shows the range of relative pressures covered by the sounds of speech. The variation among pressure ratios of different phonemes is quite apparent. The strongest phoneme in most phonetic contexts, /ɔ/ (as in f*ou*r), is 680 times more powerful than the weakest, /θ/ (as in *th*ink), in a constant phonetic environment. We know that the absolute and relative intensity of a given phoneme varies with respect to its neighbors; nevertheless, there is a degree of invariance. For example, under normal speech circumstances, the vowels are always stronger than the consonants. Within those two groups /ɔ/ , /ɑ/ , /ʌ/ will usually be stronger than /u/ , /i/ , /ɪ/ , and /ð/ , /ʃ/ , /ŋ/ will be stronger consonants than /b/ , /d/ , /p/ , /f/.

It is important to stress again that a single parameter, or measure, of the acoustic signal is a valuable tool even if it does not permit positive identification of a phoneme. It is impossible to identify a specific speech

---

*A microbar or dyne/cm^2 is the force necessary to accelerate a 1 gram mass to a rate of increase of 1 centimeter per second each second (1 cm/sec^2).

TABLE 2.1

Relative Phonetic Powers of Speech Sounds as Produced by an Average Speaker

ɔ	680	l	100	t	15
ɑ	600	ʃ	80	g	15
ʌ	510	ŋ	73	k	13
æ	490	m	52	v	12
ʊ	460	tʃ	42	ð	11
ɛ	350	n	36	b	7
u	310	dʒ	23	d	7
ɪ	260	ʒ	20	p	6
i	220	z	16	f	5
r	210	s	16	θ	1

(From Sanders, 1971, p. 48)

sound on the basis of intensity alone; because absolute intensity varies with vocal energy—it even varies from one phonetic environment to another when the overall sound pressure remains constant. We want to know to what degree a particular characteristic of an acoustic signal facilitates the correct perception of a speech sound. If we assume that we can identify any speech sound via a minimal set of acoustic characteristics, then certain *distinctive features* must be accessible to identification. The minimal amount of information is not a constant; it varies with the tightness of the constraints established. If the first few questions in the game of Twenty Questions are successfully chosen, the player may be able to predict the correct answer very quickly. The key is for the player to use each answer to constrain the field of questions, i.e., probable objects.

The constraints available within the acoustic signal function in a manner similar to the questions in the parlor game; they limit the possible choices. When enough constraints have been identified by the listener's auditory system, the identity of the acoustic pattern can be established. The intensity of the speech sound does, therefore, provide some constraining information.

As was pointed out earlier, it is possible to measure either the overall sound pressure, which includes energy across the total range of frequencies present, or the amount of energy within narrow bands centered around a particular frequency. The distribution of energy within certain frequency ranges is of considerable interest, since it conveys the pattern of acoustic energy. Since complex vibrations contain multiple frequency components each at its own energy level, frequency and intensity represent two aspects of the same vibratory event. Thus if we filter out the lowest frequencies in a speech signal, we very soon notice a decrease in the sound pressure level, indicating that the energy present at those frequencies has been removed.

The pattern of energy distribution across a specific range of frequencies constitutes a major source of constraining information about that speech signal and will be the topic of our next section.

Frequency and Frequency Structure

We have already seen that complex sound waves, such as those of speech sounds, contain vibrations of different frequencies. The overall repetition rate of the complex waveform is called the *fundamental frequency*. It is the fundamental frequency which determines the highness or lowness of the perceived pitch. *Pitch, therefore, is the psychological correlate of frequency.* In voiced speech sounds the fundamental represents the number of times the vocal folds open and close each second. Other vibrations are multiples of the fundamental: these are termed *harmonics* or overtones. Voiced speech sounds can be identified by the arrangement of their frequency components. The arrangement of the overtones for a particular voiced speech sound is, to a considerable extent, characteristic of that phoneme. Not all speech sounds have overtones that are harmonically related; the breathed or unvoiced consonants are inharmonic in structure—that is, the frequencies present do not constitute multiples of the fundamental.

We have seen that both environmental and speech sounds are complex and contain many separate frequencies comprising a complex waveform. We also noted that the frequencies present in speech sounds are arranged in various patterns which are fairly specific to each particular phoneme. This correspondence results from the relationship between the acoustic waveform and the mechanism producing the air molecule disturbance. For this reason it is not possible to divorce the entire process of acoustic reception–auditory perception from that of acoustic production. The study of acoustic phonetics necessarily involves describing and understanding the relationship between these two end stations in the total process of oral communication.

Many students have as an end goal helping children whose learning difficulties stem from auditory perceptual problems. These problems are revealed by a number of symptoms. In some children (for whom we have a multitude of labels) these difficulties will manifest themselves as problems of speech comprehension, in others through difficulty in producing correct speech, and in a third group as an inability to relate the acoustic pattern to its printed equivalent.

Thus, before we can consider the nature of the constraints imposed by frequency patterning, we should determine exactly how acoustic structuring is achieved by the speech organs. An understanding of the relationship between production and perception has become more relevant since the development of the motor theory of speech perception (to be examined later.) The hypothesis here is that, in speech perception, the listener refers to the production process in order to *restructure* the message. That is, the listener determines how the sound pattern was produced (articulated), then identifies it by its articulatory rather than its acoustic pattern.

Duration

The third characteristic of the sound wave of speech is *duration*. Like all events, the wave occupies both space and time. Gibson (1966) points out that this results in the waveform having a leading and a trailing edge, the total event constituting a *wave train*. The leading edge, which designates the *wave front*, is important in localizing the sound source. The wave train constitutes the information-bearing components essential for speech perception. Duration is, therefore, a function of the time elapse from the initiation to the cessation of the sound unit and is perceived as the length of the stimulus.

These three aspects of the acoustic wave of speech, *frequency, intensity*, and *duration*, interact in various forms to determine the particular acoustic pattern characteristics of an identifiable phoneme. In so doing they serve, progressively, to constrain us in our prediction of the linguistic value we will assign to the incoming auditory signal. This applies to the space/time boundaries of both individual phonemes and those of words and phrases. As more and more of the acoustic waveform reaches our auditory perceptual system, we operate under tighter and tighter constraints and our predictions become increasingly reliable. The nature of these constraints will be defined in greater detail in Chapter 4; at this point we need only to become familiar with those features of speech sounds that permit us to attribute a phonemic value to the acoustic signal we receive.

Speech Production

The source of the acoustic energy for speech is the respiratory system. The lungs initiate a flow of air which ultimately will leave the body by way of the nose or the mouth, thus causing turbulence in the surrounding air medium. The turbulence itself is inaudible and is not sound but merely the natural movement of air caused by normal breathing.

However, when the airflow is constrained at some location in the vocal tract, it creates an audible turbulence equivalent to the voiceless sounds or whispered vowels of speech.

For sound to be generated, the air must be vibrated; this is achieved by the action of the vocal folds. For speech production, this process involves highly structured movements that necessitate the modification or constraining of the air flow. Speech breathing differs considerably from normal respiration. Speech normally occurs on the expiratory phase of the breathing cycle. As Liberman (1972) has pointed out, this is undoubtedly because expiration is subject to far more accurate control than inspiration. Our ability to vary the duration of expiration is cited as from 3/10 second to 20 seconds; inspiration, by contrast, is a very short phase. Lenneberg has commented upon this remarkable control over a basic survival process:

> Thus it is quite clear that breathing undergoes peculiar changes during (normal) speech. What is astonishing about this is that man can tolerate these modifications for an apparently unlimited period of time without experiencing respiratory distress, . . .
>
> Our extraordinary tolerance for breathing adaptations to speech may be appreciated if we think of any other voluntary deviations from the normal breathing patterns. If we deliberately decide to breathe at some arbitrary rate, for example, faster than "normal," we at once experience the symptoms of hyperventilation (that is, lightheadedness, giddiness and other related symptoms); . . .
>
> It is fair to say that we are endowed with special physiological adaptations which enable us to sustain speech driven by expired air . . . (1967, pp. 80–81)

In conjunction with the constrained breathing patterns essential for normal speech production, major modification of the air flow is achieved within the larynx. The larynx, commonly referred to as the "voice box," constitutes a group of cartilages forming a structure which is part of the air passage from the lungs to the mouth and nose. Part of the laryngeal structure consists of two membranes, the *vocal folds*, which stretch across the air passage. The space between these two folds comprises the glottis, which can be opened and closed at will. During speech the action of the glottis serves to interrupt the steady flow of air, breaking it up into short bursts (oscillations) which form the basis, or *fundamental*, of the complex wave that reaches the listener's ear. The fundamental frequency of the speech signal is determined by the rapidity with which the vocal folds open and close. The actual range of frequencies which can be produced by the larynx is quite large, about two and one-half octaves, but normal conversation covers less than a quarter of that range. A speaker uses an habitual frequency level which is perceived as the normal pitch of his voice. The average level for adult male voices is 113 Hz, for adult females 199 Hz; for children eight years old it is 288 Hz for girls and 197 Hz for boys (McGlone 1971). This basic pitch, or fundamental, becomes the *carrier-wave* for the harmonic patterns that convey the constraint information to the listener. This basic wave form comprises the total energy of the speech signal. No further energy is added at any stage of the speech production process, but the energy represented by the harmonics will be arranged to produce the pattern necessary to evoke in the listener perception of the desired phoneme(s).

Structuring the Sound Wave

The term "structure" may be considered synonymous with pattern. Nevertheless, it is important to examine structure because of the central role played by restructuring when the listener receives a soundwave pattern.

Our discussion of speech production has not yet gone beyond the level of the larynx where the action of the vocal folds has resulted in the generation of a sound wave. This energy wave now passes into the components of the speech mechanism concerned with *resonance* and *articulation*. For us a short discussion of these processes will suffice. Should you wish to study them in depth, you will find an excellent treatment provided by Zemlin (1968), and Minifie, Hixon, and Williams (1973).

Resonance. Resonance describes a characteristic of the potential mode of vibration of air within a cavity. Each cavity has a pattern of sensitivity which centers around a particular frequency known as its resonant frequency (Figure 2.4b). This pattern of sensitivity is determined by the size, shape, and structure of the cavity. The interaction of these factors results in the air within the cavity being more sensitive to stimulus vibrations which approximate its own resonant frequency and progressively less sensitive to those which lie farther away from it. When an energy wave passes into a cavity it sets the air particles into motion. When the sensitivity at each frequency is plotted graphically we have what is known as a *line spectrum* (Figure 2.4a). If we join together the peaks of each frequency component we produce what is referred to as the *envelope* of resonant characteristics. When the stimulating sound source is itself complex, a situation arises in which two complex patterns of vibration interact. As a result of

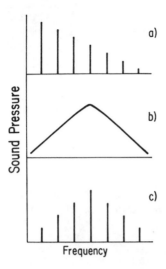

a) A line spectrum of a sound.

b) The envelope of the resonant sensitivity of a cavity.

c) The rearranged frequency components after being resonated.

Figure 2.4   The effect of resonance on a sound. (Sanders, 1971, p. 45)

this action we find that the harmonic pattern of the stimulus wave becomes modified; the component harmonics are rearranged (resonated or "resounded") to conform to the envelope of resonant characteristics (Figure 2.4c). Regardless of the fundamental vibration rate of the stimulating sound wave, the air within the resonating cavity will vibrate at its own resonant frequency. How often it vibrates thus will be determined by the rate of vibrations of the stimulus energy. So a compromise occurs; the *pattern* of vibrations is determined by the resonance characteristics of the cavity, but the *rate* at which this pattern repeats itself is determined by the vibration rate of the stimulus sound. *The repetition rate, or fundamental frequency, of the stimulus sound determines the perceived pitch of the resultant sound: the pattern of resonance determines its perceived character and quality.*

Now let us look a little more closely at the manner in which the rearrangement of the harmonic vibrations is made. We saw earlier that the energy wave of speech at the level of the larynx is complex; it contains a fundamental frequency and has harmonics which are multiples of that frequency. If we compare the pattern of energy distribution of the vocal tone with the sensitivity pattern of a particular resonance chamber, or cavity, we will see that at some frequencies there will be coincidence of harmonics, but that at others the two patterns will not coincide. When the energy represented by the harmonics of the two sound waves exactly coincides, that particular harmonic is strengthened; when the two energy components are completely noncomplementary the vocal harmonic is suppressed, or, more correctly, *damped*. Thus the energy distribution within the vocal tone is rearranged through the process of resonance. It has been explained that each cavity will have a particular resonant quality determined by its shape, size, and composition. Speech production demands that a wide range of qualities or characteristics be impressed upon the vocal tone. This can be achieved either by the use of a large number of resonators or, as is actually the case, by varying the shape and size of a few resonators.

The passageway from the larynx to the atmosphere consists of a series of *coupled resonators*, (linked cavities) which include the pharynx with its laryngeal, oral, and nasal sections, the mouth, or oral cavity, and the nose. There are also several secondary resonators, particularly the sinuses, but the contribution they make to speech perception is small so we shall not consider them.

The primary resonators are able to change their resonant spectra by altering (1) their volume or (2) their shape. The organs which effect these changes are known as *articulators*, and the process is referred to as speech articulation. The organs involved consist of the lower jaw, which modifies the volume of the oral cavity by raising or lowering; the lips, which together with the cheeks, vary the size and shape of the external opening and the size and shape of the oral cavity, thus changing its volume; the soft palate, which raises and lowers, effectively separating the oral and nasal cavities when it is raised against the rear wall of the naso-pharynx; and the tongue, which, by its varied placement within the oral cavity, changes the cavity shape and volume.

Each time one or more of the articulators moves, the resonant spectrum of the coupled resonance cavities is changed. As a result, the glottal energy can be structured to assume a wide range of different patterns, each of which is characteristic of the particular arrangement of the articulators at the time the energy passes through the system. This relationship between articulator placement and the resultant acoustic structure is highly important, since it provides a valuable clue to the possible mode of signal decoding performed by the auditory system of the listener.

Let us now review what we have considered to this point. We have seen that energy originating in the lungs passes into the larynx where the vocal folds are voluntarily caused to vibrate. The vibration of the vocal folds in turn sets the surrounding air molecules into motion initiating the sound wave. The rate of vibration of the vocal folds determines the fundamental frequency of the complex glottal tone, which is perceived as the pitch of the voice. Although the vocal pitch fluctuates during speech, it tends to center around a particular frequency for each speaker. This frequency, which is referred to as the speaker's *natural pitch*, is, by virtue of the size of the larynx, higher in women and children than it is in men. The complex vocal tone enters the resonators of the pharynx, mouth, and nose. Here the harmonics of the vocal tone are patterned, or structured, according to the degree of coincidence between the vocal tone harmonics and the resonant characteristics of the cavities. Frequencies which closely approximate are reinforced, those which lie farther apart are damped. Thus, by changing the shape and size of the resonators it is possible to create and recreate at will the range of specific speech sound patterns that form the basis of oral communication.

Let us examine now the nature of the rearrangement of the energy distribution resulting from resonance.

Ironically our understanding of the nature of the acoustic signal, and of the auditory system, has been achieved by use of our visual system. What we know about acoustic phonetics we have learned, to a great extent, from graphic representation of the structure and changing nature of the sound waves of speech. So we use our eyes to understand our ears. It is also important to remember that our understanding is limited, expanded, and defined by the nature of the technology available to us. We cannot escape the role technology plays in determining the ways in which we come to understand the nature of the physical environment.

The study of the physical nature of speech sounds was for many years severely hampered by the absence of appropriate instrumentation to permit the rapid analysis and graphic display of the components of the sound wave. Fischer-Jorgensen has stated:

> The old instrumental phonetics was mainly concerned with the analysis of the physiological aspect of speech sounds; it was not that there was no interest in the acoustic aspect, but the means of analysis which the phoneticians of the nineteenth century had at their disposal did not allow them to proceed very far along these lines. The technical developments of this century opened new possibilities already before the war, and various phoneticians availed themselves of these new possibilities (e.g., Gemelli, E. Z. Wirner), but it is (sic) not until the last decade that there has been an almost explosive development in the field of laboratory acoustic equipment. A number of new machines can now be utilized in the analysis of speech sounds, or have been specifically constructed for this purpose. (1961, p. 114)

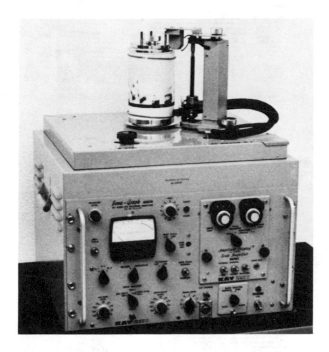

Figure 2.5   A sound spectrograph.

There is little doubt that the machine which opened up the greatest opportunities for analyzing and defining the acoustic structure of speech sounds was the sound spectograph (Figure 2.5). Since we shall be using spectrograms to study the speech signal, we shall review in some detail the operation of this widely used tool.

The spectrograph

The term *spectrum*, or in the plural *spectra*, refers to a full range of frequencies. Commonly used in reference to the distribution of light energy, it is also used to describe the distribution of sound energy across a range of frequency vibrations. The minimal and maximal frequency components of speech fall within a band from approximately 100 to 8000 Hz. The spectrograph provides a spectral analysis (spectrogram) of the energy present at each frequency or band of frequencies within a complex acoustic signal (Figure 2.6).

The spectrogram depicts the frequency range on the vertical scale and the time on the horizontal scale. The intensity of the energy present within any band of frequencies is shown by the darkness of the marking; the more energy present within a band of frequencies, the blacker the plot. In this

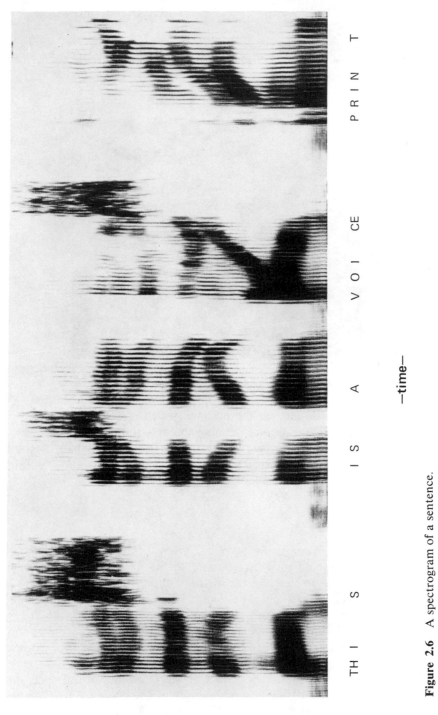

Figure 2.6 A spectrogram of a sentence.

way the relative intensities within various frequency bands can be observed for a single sound, or they may be compared between sounds. Thus the three critical parameters of the speech sound, frequency, intensity, and duration, are simultaneously displayed on the spectrogram.

The process of producing the spectrogram first involves the tape recording of the sample onto a tape-loop within the instrument. This permits the signal to be repeated exactly over and over again during its analysis. The second stage involves passing the sound energy through a bank of filters which are divided into frequency bands. Either a wide band (300 Hz) or a narow band (45 Hz) setting may be selected. The narow band filter records each harmonic separately. As the interrogating band moves across the frequency range from low to high it permits the flow of an electrical current to a "pen" which electrically arcs a spark through the recording paper on the drum, charring the surface with a density proportional to the energy in the acoustic signal: the greater the acoustic energy, the stronger the electrical current and, therefore, the blacker the marking made by the pen. In this way, the energy in a complex wave is broken down and distributed according to the individual frequencies that originally combined to produce the complex wave.

We have already seen that the harmonic pattern of the laryngeal tone is rearranged by the supralaryngeal system of coupled resonators and articulators. We saw how the acoustic pattern can be modified at will by rearranging the articulators. The effects of these modifications, as may be expected, are to be seen in the spectrogram. The spectrogram for a male producing the vowel sound /a/ is shown in Figure 2.7. From this figure we can observe that the articulatory-posture characteristic of /a/ has resulted in the concentration of the acoustic energy into clearly identifiable energy bands. These concentrations are called *formants* and are present in all vowels and voiced consonants. They are numbered sequentially from F1, the lowest, to F3, the highest audible speech formant, above which no measurable contribution to phoneme identification is made. Thus for the /a/ sound we have F1, which centers around a frequency of 700 Hz, F2 with a mid-frequency of approximately 1100 Hz, while F3 concentrates around 2400 Hz.

This acoustic pattern results from the passage of the glottal energy through the articulatory-resonant system when it has been placed in the particular posture associated with the articulation of /a/. The acoustic pattern can therefore be considered to be representative of what is occurring in the speaker's mouth. *The acoustic event possesses equivalency to the articulatory event.* Research studies have shown fairly conclusively that F1 and F2 are critical to the perception of the speech sound. While F1 and F2 contribute to the identification of the *phonemic* category to which the

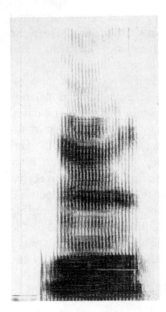

/a/

Figure 2.7 A spectrogram of a male voice saying /a/.

sound belongs (i.e., whether it is an /ɪ/ /i/ or /u/ , or /z/ /v/ or /ʒ/ ,)F3 has been shown to contribute both to the perception of labial, alveolar, and velar stops as well as to the *allophonic* qualities of the phoneme (Liberman, Delattre, Cooper, and Gerstman, 1954). Allophones are perceptually discriminable variations of a given phoneme.

For the same speaker, or for groups of males, females, or children, the center frequency of each formant is a fairly consistent figure. This is not true, however, when comparisons are made between groups. The difference between the frequency figures for men and women have been shown to differ by as much as 17 percent.

Fischer-Jorgensen (1961, p. 122) has concluded that the relations between the formants (the difference figure between the two mid-frequencies) is to some extent decisive for perception of the phoneme. Petersen and Barney (1952) and Dunn (1950) have demonstrated that these differences are related to the varying mean size of the oral cavities in men, women, and children. Apparently it is not necessary for the listener's brain to handle three distinct sets of perceptual reference tables for the various groups of speakers, nor need he have subsets for individual variations within the group. Ladefoged and Broadbent (1957) have suggested that a listener applies a conversion process to the incoming acoustic signal. This adjusts his perceptual system to the speaker's speech pattern by means of a correction factor computed from a rapid analysis of the first few utterances of the speaker. This can be envisaged as a tuning process which adjusts, or tunes, the incoming auditory pattern to compatibility with one's own internal pattern, a process referred to as *normalizing*. In this way we allow for, and therefore eliminate, the difference between the speaker's articulatory pattern and our own. We are thus able to reduce the impedance, or mismatch, between the two perceptual systems. You will probably have noticed on occasions when you have spoken with a person whose speech is initially somewhat difficult to follow, that it appears to become progressively easier to understand as he converses with you. We say that we "become used to his manner of speaking," a recognition of the internal adjustment our perceptual system has made. In fact, if we are continually in contact with this person's speech, the perceived dialect begins to fade. For example, my British dialect is quite unapparent to my wife and children, but readily detected by strangers.

These examples add support to the contention that we hear others in our own translation, or by reference to our own system. If we can support this claim, we may gain considerable insight into the auditory perceptual difficulties encountered by many children who have difficulty in achieving normal speech articulation, or stumble over phonics in learning to read. We shall examine this proposition in Chapters 9 and 10.

The relative, rather than the absolute, position of the formants appears to make a major contribution to the identification of the speech sound. The formants serve as an additional source of constraint by combining frequency, intensity, and duration in a particular manner. Through the process of articulation and resonance we imprint certain patterns on the glottal wave which become characteristic of specific phonemic values. This can be seen from the four spectrograms of the vowels /i/ and /o/, and the consonants /d/ and /g/ shown in Figure 2.8.

Acoustic Characteristics of Consonants

The physical cues to the phonemic identification of vowels appears to involve primarily the recognition of patterns of formant relationships which vary by the parameters of frequency, intensity, and duration.

The perception of consonants is also in part dependent upon the arrangement of the acoustic energy. However, the consonant spectrum is not analyzed primarily in terms of the differentiation of specific phonemic acoustic patterns but in terms of the manner of their production. Liberman

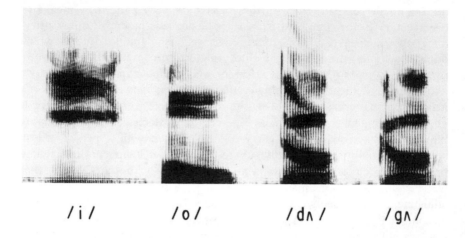

/i/   /o/   /dʌ/   /gʌ/

Figure 2.8 Spectrograms of the vowels /i/ /o/ and the consonants /d/ /g/.

(1957) has suggested that, for the sake of convenience, consonant cues be divided into three articulatory groups. The classes he has proposed are:

1. cues arising from constriction of the air flowing during production of the sound
2. cues arising from the presence or absence of nasalization
3. cues arising from articulatory movements

The information derived from measurement of constriction, movement, and nasalization or its absence act as additional constraints to more accurately define the pattern. Remember that our mind must utilize the acoustic cues to reconstruct the message. Thus the greater the number of available cues, the greater the probability of correct identification.

Let us look at the three classes of consonant cues suggested by Liberman.

(1) *Constriction cues:* These arise from consonants whose production is dependent upon friction, partial friction (affriction), or plosion. They include the phonemes /f/, /v/, /θ/, /ð/, /s/, /z/, /ʃ/, /ʒ/, /tʃ/, /dʒ/, /p/, /b/, /t/, /d/, /k/, /g/. The acoustic characteristic of these constriction sounds is their short duration; they do not greatly influence the adjoining phoneme since they rapidly fade as the articulators open toward the next phone (sound). Among the cues shown to be important are the frequency of the friction noise, particularly the lower frequency limit which facilitates the differentiation between /s/ and /ʃ/ and to some extent between /f/ and /θ/ (Harris, 1954), and the frequency of the air released in the pro-

duction of the voiceless plosives /p/ , /t/ , /k/. Similar cues exist in the data concerning the sound onset and duration. The important point is that the perceived presence or absence of friction noise either rules in or rules out the whole category of fricative consonants, considerably narrowing the field of possible phonemes.

(2) *Nasality cues:* The second type of cue lies in presence of the nasality characteristic of the consonant group /m/n/ŋ/. Presence of the nasality cue clearly limits the perception to one of these three consonants, ruling out all others.

(3) *Transitional cues:* The third source of constraining information has to do with the fact that the rapidity of speech makes the precise articulation of each speech sound impossible. In fact, if we establish an optimal articulatory position for a particular speech sound, that is to say, an ideal target description, we observe that on repeated productions of that speech sound, the target criteria are seldom, if ever, exactly fulfilled. The articulatory, and therefore the acoustic, properties of the speech sound vary slightly from one production to the next within the same speaker. The perceptual system's ability to adjust to these approximations is, as we have already pointed out, critical to the perception of speech at the rate occurring in normal speech communication. Thus the linking of speech sounds involves a process resembling the movements of a slalom skier, skirting but not actually hitting each marker. In speech production our articulatory-resonant system slaloms from one target position to a very close approximation of the next phoneme target, but never exactly reaches it. Like the skier, before we arrive at the target we are already veering away from it in order to head toward the next. If we wish to study the progress of the downhill skier, we could look at a series of still photographs each representing his position at the target area. We might refer to these as the "static" positions, illustrative of each flag position, i.e., "steady-state photos." Yet it is quite clear that this series does not in fact provide us with all the details necessary to fully analyze a skier's run. To do this we turn to the movie film which permits us to observe how he approached the target position and how he moved away from it toward the next. We would see from the series of photos which make up the movie film, that each target position was approached differently depending upon its relationship to the previous target. Indeed, if we were to study the approach to one gate or target position for varying amounts of snow banking and snow conditions, we would observe that the skier's approach to the same optimal gate would vary on each run.

This situation is analogous to that of speech production. As the articulatory organs move from a consonantal position to the next phone (sound), they generate an intermediate patterning of the glottal energy, which shows up in the acoustic wave as frequency shifts preceding the formants. These linking characteristics are termed *transitions*, since they are the result of changing from one target position to the next. Liberman has stated:

> We know that these transitions are not merely the incidental acoustic accompaniments of the movements that a speaker must make when he goes from "consonant" to "vowel." Rather, they are perceptual cues, and it is difficult to exaggerate their importance. (1957, p. 118)

Stetson (1961) in developing a theory of speech production based upon the ballistic movements of the articulators suggested that consonants do not exist in their own right but serve only as vowel modifiers. When you realize that, with the exception of the continuants and nasals, a consonant cannot be produced except with a vowel, the claim is convincing even without study of his rationale.

Regardless of whether or not the consonant exists as a discrete acoustic element, it does exert a distinctive influence upon the adjoining vowel. The particular vowel modification depends upon the consonant in question: Each consonant will affect each vowel differently, creating a wide range of transitional patterns, each identifiable with a consonant-vowel, vowel-consonant relationship, and therefore representative of that particular phoneme combination.

The following spectrograms illustrate these transitional changes. First, compare the formant structure for the vowels /i/, /a/, /u/ shown in Figure 2.9. The mid-formant values for these three vowels are:

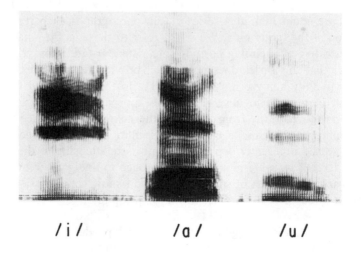

/ i /   / a /   / u /

Figure 2.9   A spectrogram of the vowels /i/, /a/, /u/.

	F1	F2	F3
/i/	300	2300	3250
/ɑ/	700	1200	2500
/u/	300	900	2250

There are clearly definite differences between the loci of each formant for the three vowels. The difference in frequency position for F1 and F2 are therefore:

	F2 minus F1
/i/	2000 Hz
/ɑ/	500 Hz
/u/	600 Hz

Now consider the acoustic effect of preceding the vowel /i/ with the voiced stop (or plosive) consonant /d/ (Figure 2.10). You will observe that the formants now seem to have a tail, which in the graphic depiction trails behind them. In reality this tail precedes the vowel; it constitutes the "transition" from the preceding consonant. You can see that this occurs in both F1 and F2. It also occurs in F3, but we have agreed to concern ourselves only with the first two formants. The general configuration, or pattern, of change is a rising frequency before both formants.

By contrast, the transitional configuration for /du/ is a falling pattern. Spectrograms for /bi/, /ba/, /bu/, /di/, /da/, /du/; and /gi/, ga/, /gu/ are shown in Figure 2.10.

These figures show how the acoustic changes resulting from the movement of the articulators from one posture to the next appear on the spectrograph. They illustrate how the transitions follow different patterns as a function of the relationship between the consonant and the vowel. It is possible to depict these transitions more graphically by painting the formant transitions on a frequency/time chart. This procedure is used in preparing artificial spectrograms for synthesizing speech. Speech synthesis for the psychophysical study of speech is becoming a very sophisticated approach to the study of the relationship of the physical acoustic signal to the perceived linguistic value. It is now perfectly possible to synthesize intelligible sentences and to produce artificial vowels, and some consonants, which are indistinguishable from human speech sounds. The artificial spectrograms are painted in a white paint and fed through a playback machine which transduces light energy reflected by the paint patterns into patterns of sound energy (Delattre, Cooper, Liberman, and Gerstman, 1952). The procedure permits identification of patterns of transitions, which will evoke the perception of a particular phoneme before a particular vowel. Figure 2.11,

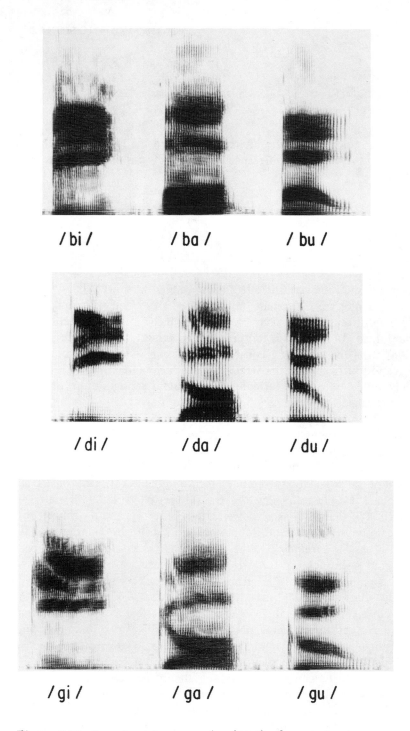

/ bi /   / ba /   / bu /

/ di /   / da /   / du /

/ gi /   / ga /   / gu /

Figure 2.10 Sound spectrograms showing the format structure of the vowels /i/ , /a/ , /u/ in isolation (Figure 2.9) and when preceded by the consonants /b/ , /d/ , and /g/.

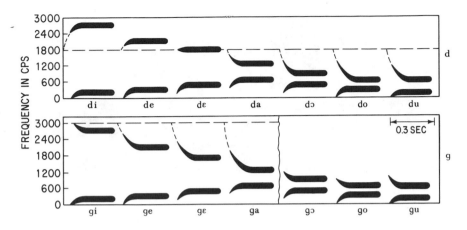

Figure 2.11 Spectrographic patterns that produce /d/ and /g/ before various vowels. The dashed lines are extrapolations to the /d/ and /g/ loci. (Liberman, 1957)

taken from Liberman's work (1957), shows these transition patterns for /d/ and /g/ before seven vowels.

This structuring of the acoustic wave at the level of the phoneme is perhaps most clearly seen when we examine the spectrogram for a sentence (Figure 2.12). The important point is that the transient changes observable in the spectrum, together with the stable, or *steady state*, components of the vowels, each with its distinctive formant structure, constitute acoustic information constraining the listener's choice of possible phonemes.

Intonation

There remains one aspect of the acoustic speech signal to be considered, namely intonation. Lieberman (1967a) has stated that intonation has commonly been assumed to convey only the speaker's emotions or his attitudes toward the words of his sentences. He believes, however, that intonation plays a central role in the process of recognition of syntactical arrangement of words. In his opinion, intonation provides acoustic cues that segment the speech signal into linguistic units for syntactic analysis. In a later publication (Lieberman, 1972) he discusses the manner in which the intonational information is impressed upon the acoustic waveform. According to his text, the intonational contours are carried by the fundamental frequency, which in turn is controlled by variation of the air pressure in the laryngeal tract before it reaches the glottis (subglottal pressure) and by changing the tension of certain laryngeal muscles. The discussion of exactly how this mechanism works is beyond the scope of this text; in-

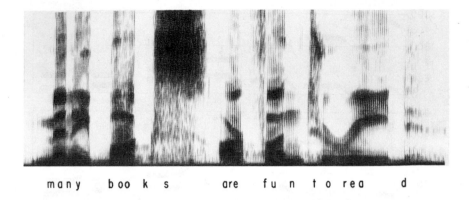

many   b oo k   s   are   fu n   t o rea   d

Figure 2.12   Sound spectrogram of the sentence: *Many books are fun to read.*

terested readers are referred to the original publication. It is nevertheless a significant concern since intonation is known to influence the perceived meaning of a sentence and may, therefore, prove a critical constraint factor derived from the acoustic waveform.

It was speculated that intonation might operate through specific sound patterns referred to as "pitch morphemes," each having a distinct meaning, but this approach proved unsuccessful. Lieberman (1967a,b), took a different approach. He suggested (1967 b, p. 314) that "the primary linguistic function of intonation is to furnish acoustic cues that allow the listener to segment speech into blocks for syntactic processing." In other words, the intonational cues serve to constrain us in our perception of the relationship between the words we then restructure from the acoustic signal. Only when we have restructured the intended grouping of words (phrasing), are we able to attribute semantic value, or meaning. The phrase marker, derived from the acoustic wave and from the listener's knowledge of the rules of the language code, is critical to correct perception of the intended meaning.

> Intonation furnishes acoustic cues that tell the listener when he has a block of speech that constitutes a satisfactory input to his syntactic recognition routines. Intonation can furnish different meanings to utterances that have the same words by grouping the words into different blocks which direct the listeners' recognition routines toward one underlying phrase marker rather than another. (1967, p. 315)

REFERENCES

DELATTRE, P., F. S. COOPER, A. M. LIBERMAN, AND L. L. GERSTMAN, 1952. Synthesis as a research technique. Proceedings of the VIIth International Congr. of Linguists, London.

DUNN, H. K., 1950. The calculation of vowel resonance and an electrical vocal tract. *J. Acoust. Soc. Amer.*, **22**, 741–53.

FISHER-JORGENSEN, E., 1961. "What can the New Techniques of Acoustic Phonetics Contribute to Linguistics?" in *Psycholinguistics*, ed. S. Saporta. New York: Holt, Rinehart, and Winston, p. 114.

GIBSON, J. J., 1966. *The Senses Considered as Perceptual Systems.* Boston: Houghton Mifflin Company.

HARRIS, K. S., 1954. Cues for the identification of the fricatives of American English. *J. Acoust. Soc. Amer.*, **26**, 952 (a).

LADEFOGED, ·P., 1962. *Elements of Acoustic Phonetics.* Chicago: University of Chicago Press.

LADEFOGED, P., AND D. E. BROADBENT, 1957. Information conveyed by vowels. *J. Acoust. Soc. Amer.*, **29**, 98–104.

LENNEBERG, E. H., 1967. *The Biological Foundations of Language.* New York: Wiley.

LIBERMAN, A. M., 1957. Some results of research on speech perception. *J. Acoust. Soc. Amer.*, **29**, 117–23

LIBERMAN, A. M., 1961. Some results of research on speech perception, in *Psycholinguistics*, ed., S. Saporta. New York: Holt, Rinehart and Winston, pp. 142–52.

LIBERMAN, A. M., P. C. DELATTRE, AND F. S. COOPER, 1954. The role of consonant-vowel transitions in the perception of the stop and nasal consonants. *Psychol. Monographs*, **68**, 1–13.

LIEBERMAN, P. 1967(a). *Intonation Perception and Language.* Research Monograph **38**. Cambridge, Mass.: The M.I.T. Press.

LIEBERMAN, P. 1967(b). Intonation and the syntactic processing of speech, in *Models for the Perception of Speech and Visual Form*, ed., W. Wathen-Dunn. Cambridge, Mass.: The M.I.T. Press.

LIEBERMAN, P., 1972. *Speech Acoustics and Perception,* in The Bobbs-Merrill Studies in Communication Disorders, ed., Halpern. New York: Bobbs-Merrill.

McGLONE, R. E., 1971. The Acoustic Aspect of Speech, in D. A. Sanders, *Aural Rehabilitation.* Englewood Cliffs: Prentice-Hall.

MINIFIE, F. D., T. J. HIXON, AND F. WILLIAMS eds., 1973. *Normal Aspects of Speech, Hearing, and Language.* Englewood Cliffs, N.J.: Prentice-Hall.

PETERSEN, G. AND H. BARNEY, 1952. Control methods used in a study of the vowels. *J. Acous. Soc. Amer.*, **24**, 175–84.

SANDERS, D., 1971. *Aural Rehabilitation.* Englewood Cliffs, N.J.: Prentice-Hall.

STETSON, R. H., 1961. *Motor Phonetics.* Amsterdam: North Holland Publishing Co.

ZEMLIN, W. R., 1968. *Speech and Hearing Science: Anatomy and Physiology.* Englewood Cliffs: Prentice-Hall.

3

The Auditory System—
A Pattern Processor

"The ear is the organ of hearing." This statement represents our traditional understanding of the auditory system, yet it is both imprecise and incomplete in defining the nature of the system we use in hearing. From our discussion in the previous chapter it will be apparent that the process of hearing must be highly complex to be capable of handling the enormous range of sound pressures, the complexity of acoustic signals, and the many difficult listening situations we encounter. It will be our task to attempt to become more familiar with this process.

Whenever we wish to understand something, whether it be an object, a device, a system, or a process, we begin by carefully describing what can be observed. The anatomical study of the ear has been concerned with such careful observation and description. It has provided us with a highly developed knowledge of the structure of the auditory system. Our understanding has been furthered enormously by the development first of the microscope and, subsequently, of the operating microscope, which has made possible the dramatic surgical achievements of the otologist. More recently the electron microscope has permitted the anatomist to observe in detail even smaller structures. Technical advances have made it possible to slice extremely thin sections of the temporal bone and the sensory end-organ of hearing contained within it, and to stain the specimens so that they can be read with increasing clarity, adding to the study of the deep structures of the auditory system. Finally, the establishment of temporal bone banks in several centers in the United States is leading to the comparative anatomical study of normal and defective ears.

We need to keep in mind that technological sophistication and the understanding it lends to anatomical structures do not provide the complete answer to how we perceive sounds. The perception of sound involves *listening*, a far more complex process than hearing. Nevertheless, the auditory system itself has provided us with some important clues from which current theories of hearing and perception have been evolved.

If we examine a diagram of the human ear in any standard textbook, we will notice that a single ear is depicted. Of course, it is unnecessary to point out that an identical structure exists on the other side of the head. Nevertheless it is important to recognize that, as a perceptual system, the auditory pathway is binaural. The existence of two ears insures a process more complex than that which occurs when a monaural system exists. A binaural system provides the basis for comparison of two sets of data pertaining to the same stimulus. Furthermore, this binaural system, part of the human head, is mounted on the neck, making the whole system a mobile one. This mobility is important because it contributes greatly to the role of the auditory system as an active one, which seeks out information from the enormous array of stimuli generated by the changing environment.

Many of us are familiar with the basic structures of the ear. We shall reconsider them, however, to emphasize the roles they play in the transmission and analysis of acoustic information—information that permits the restructuring of a thought or event as a perceptual experience.

We have seen that objects (including people) and events broadcast information by changing the media surrounding them. Speech production is based upon the ability of the individual to pattern the movement of the air particles around him by manipulation of the vocal-articulatory system. In this way information in the form of patterns of air pressure changes is broadcast into the air medium. The actual amounts of pressure change are exceedingly small. For this reason the auditory system requires a highly sensitive mechanism for detecting these small variations in acoustic pressure and for conveying the energy patterns to the sensory end-organ of hearing where they will be internalized into the sensori-perceptual system. It should be understood that the real barrier between our perceptual system and the outside world is at the level of the sensory end-organs themselves. The end-organ is the component of a perceptual system responsible for internalizing information patterns. It does this by recording changes in the stimulus event in the form of a neural code, making the information available to the nervous system for analysis and interpretation. This process occurs at the retina of the eyes, at the taste buds of the tongue, and at the hair cells of the organ of hearing. Although the auditory system extends to the mechanical components of the middle and inner ear, the process of perception can begin only when the information pattern has entered the nervous system.

The External Ear

The Auricle.   When we look at an ear in a lay capacity we observe the *auricle* or *pinna*, which is the flap of cartilage and skin attached to the side of the head. Although some of us are blessed with the doubtful attribute of being able to "wiggle" our ears, the auricle can legitimately be considered a nonmobile component in humans. Such is not the case with the rabbit who, along with many other creatures, may be seen listening carefully, searching for the sound which has caused him to freeze. He does so by turning the auricles independently almost a full circle. The part played by the human auricle in nonauditory functions cannot be denied, since it constitutes an important erogenous zone. Its role in auditory perception, by contrast, is far less important. It has been well established that it does little if anything to enhance the perceived loudness of a sound, and has generally been considered to have no influence on auditory behavior. However, writing of the mechanical properties of the ear, Békésy and Rosenblith (1958) remind us that experiments carried out in the late 1800s indicated that, although the auricle does not contribute to the preservation of the intensity of sound energy in the speech range as it enters the ear canal, it apparently does make a contribution to the individual's ability to localize sound. This is particularly so when the sound originates from immediately in front of or behind the listener. Recent experiments have added support to this view (Batteau, 1967; Bauer et al., 1966; Freedman and Fischer, 1968).

The Ear Canal.   From approximately the center of the auricle, often concealed by a cartilaginous projection (the tragus), runs the ear canal or *external auditory meatus.* This is a tube which becomes narrower as it goes deeper into the side of the head. It serves the important role of providing a channel by which the sound-wave energy can reach the eardrum stretched across its internal bony walls.

So far, the sound energy has not impinged in any way upon the auditory mechanism. It has entered the external ear canal and has been slightly modified by the canal, which, like any cavity, has a broadly tuned resonant sensitivity. Its resonant frequency centers around 3800 Hz (Zemlin, 1968, p. 367). This resonance serves to enhance those frequencies in the complex waveform which approximate the resonant frequency. As such it plays a small part in the total tuning process of the hearing apparatus to provide greatest sensitivity to those frequencies within the normal range of human speech sounds.

The auricle and the external auditory meatus together constitute what is known as the *external ear*, whose purpose is simply to channel the sound wave energy to the eardrum, or *tympanic membrane*, the initial structure of the middle ear.

The Middle Ear

The Tympanic Membrane.   This membrane represents the first component of the mechanical system of the ear. The system has evolved for the purpose of conveying the information contained in, or impressed upon, the energy wave, from the air medium to the fluid-filled end-organ of hearing. Topographically, the tympanic membrane serves as a thin membranous wall dividing the outer ear from the drum-like cavity of the middle ear. The membrane is cone-shaped with a thickened border by which it is attached to a circular groove in the bony wall of the ear canal (Figure 3.1).

Before we consider its structure, we should remember the purpose the drum must serve. At this point the system needs a way of replicating the pattern of constraint information impressed upon the acoustic energy wave. That pattern still exists as movements of air molecules creating pressure changes (compressions and rarefactions) in the air present in the ear canal, and must reach the organ of hearing where it will be internally represented within the auditory perceptual system. The tympanic membrane is the first component of a mechanical chain which will permit the energy pattern to reach the sense organ of hearing. In order for the membrane to duplicate a pattern of pressure changes, it must be sensitive to them and be moved by them. To achieve this purpose the tympanic membrane has evolved as three layers of tissue. The most important layer is the middle fibrous layer which is sandwiched between an outer layer of skin and an inner layer of mucous membrane. It is the fibrous layer which provides the elastic properties necessary for the drum membrane to stretch and contract as the pressure impinging upon it varies. The fibers of this layer run in two directions to form what appears to be a woven matrix with

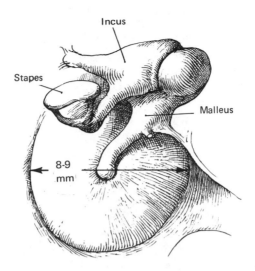

Figure 3.1   The eardrum membrane.

some fibers running in circular direction while others radiate from the center of the cone to the edge. The fibers are not actually interwoven, but consist of closely connected layers. The greatest portion of the membrane is tense like the membrane of a drum; this part is consequently referred to as the *pars tensa*. A small portion of the upper drum area is much less taut because of the presence of less fibrous tissue; this is referred to as the *pars flaccida*. The eardrum comprises a structure which may be vibrated by the changing pressure waves funneled onto it along the ear canal. This vibration results from the composition and tension of the drum membrane. The tympanic membrane constitutes, therefore, an important component in the mechanical system, since it is capable of receiving, with little energy loss, the information-bearing acoustic wave. The qualification of "with little energy loss" is a particularly important one, since the pressure changes to be responded to are so very small. If there is any significant reduction in the amount of acoustic energy flowing from one medium to another, the task of extracting the information content becomes increasingly difficult. The opposition a system presents to the flow of energy through it is known as *impedance*. When the impedance of a system is raised, as in a mechanical or conductive hearing loss, it is more difficult to transmit energy through the system.

Impedance.   The potential to change energy from one form to another (transduction), and to pass it from one system to another with little loss of energy, is a function of the degree of compatibility between systems.

The acoustic impedance of a medium is the opposition of that medium to the flow of acoustic energy through it. Compatibility between systems depends upon the degree of elasticity and the density of the mass of each system's components. When these two characteristics differ considerably between systems, the systems are said to be mismatched and, therefore, not compatible. If two systems are highly compatible, the energy, ergo the information content, will flow easily from one to the other; that is to say, *the impedance will be low*. The acoustic impedance of a medium, of course, varies inversely with the degree of compatibility. If two systems are barely compatible, the impedance will be high, and very little energy will be transduced.

The major function of the middle ear is to serve as an impedance-matching device between the outer air and the fluids of the sensory organ of hearing. Without this structure, most of the useful information in the speech signal would not be available to the listener because of the impedance mismatch between the air and the cochlear fluids which activate the sensory organ of hearing.

The Cavity of the Middle Ear. If we were able to stand on the outside of the eardrum and peer through it, we would observe on the other side of it a small, air-filled space, 1 to 2 cubic centimeters in volume. This is the middle ear cavity or, as it is correctly called, *the tympanic cavity*. Figure 3.2. It is a vaulted cavity, almost four times as high as it is long. The approximate dimensions of it are shown in Figure 3.3. Directly opposite the tympanic membrane is an oval shaped window in the far wall of the cavity. This wall divides the middle ear cavity from the inner ear cavity on the other side of it; the *oval window* provides communication through the wall. Beneath the oval window the wall bulges into the cavity in the form of a promontory, and beneath this *promontory* there is another window appropriately termed the *round window*. This window also communicates between the middle and inner ear. We will talk about its function when we discuss the inner ear.

From the tympanic cavity, a tunnel runs downward through the bony floor of the middle ear cavity. This is the internal end of the *eustachian tube*. If we were able to travel down it, we would find that the bony section ends and is replaced first by a cartilaginous and finally by a membranous section. The cartilaginous section of the tube is normally in a collapsed state, opening and closing when we swallow or yawn. Should we pursue our travels down the tube, we would eventually find ourselves in the pharyngeal part of the throat at the level of the nose (nasopharynx).

There is a direct relationship between the eustachian tube and the tympanic membrane. We have pointed out that the absolute pressure

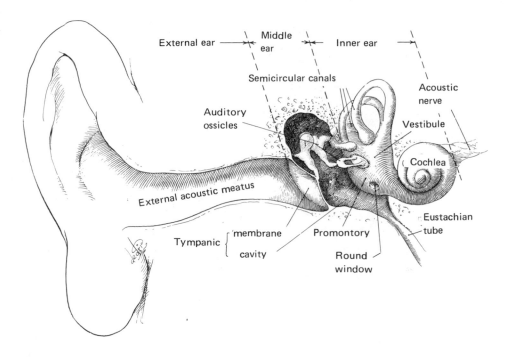

Figure 3.2   Drawing of the middle ear showing its relationship to the inner ear.

changes occurring at the eardrum during speech are extremely small. The
membrane has evolved to a degree of tension which makes it highly sensitive to these small changes. If the system is to provide the optimum sensitivity, the natural tension must be maintained. This means that the static
pressure of air on either side of the tympanic membrane must be equal
when the membrane is at rest. It is the role of the eustachian tube to
ventilate the tympanic cavity, equalizing the pressure on both sides of the
drum. If this function of the tube is impaired, as it is in certain patholog-
ical conditions of the middle ear, the tympanic cavity actually becomes a
closed space. The air in that cavity, first the oxygen, then the nitrogen,
now begins to be absorbed by the mucous membrane lining, causing a
progressive reduction of the air pressure in the cavity. When this occurs,
the pressure on the external side of the tympanic membrane is higher than
that on the inside; this results in the drum's being pushed inward, or *re-
tracted*, causing an abnormal tension to be placed on it. This retraction
reduces the sensitivity of the system because some of the information-bearing acoustic energy is consumed in overcoming the increased resistance
of the drum, and so does not enter the auditory system. In other types of
ear pathology, fluid or pus may fill up the middle ear causing the drum to

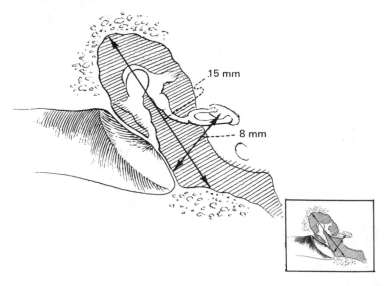

Figure 3.3 Approximate dimensions of the tympanic cavity of the middle ear.

be pushed outward, or *distended*. The end result is the same: increased tension on the drum increases its impedance, thus reducing signal energy and leading to a loss of potential information.

The enormous impedance represented by the space between the tympanic membrane and the oval window opposite it must be overcome. Sound traveling across this space with no intervening impedance-matching mechanism loses approximately 30 dB SPL, which reduces a loud voice to the level of a soft one. This amount of energy loss would make conversational speech considerably less audible by consuming acoustic energy and its information content that should have entered the auditory processing system.

Fortunately there has evolved in the human ear a chain of three small bones, the *ossicular chain*, connecting the tympanic membrane with the membrane of the oval window. Herein lies the solution to the problem of energy conservation. These bones, the *ossicles* of the middle ear, articulate with each other in such a way that a movement of one causes a resultant and equivalent movement of the other two. In this way they permit the information-bearing energy pattern to flow through them with minimal impedance.

The Ossicles. To examine the ossicles of the human ear is to experience sculptured perfection in miniature. These are the smallest bones in the human body, yet each is exquisitely shaped, and each relates perfectly with its neighbor. They have been named after objects which they were felt to resemble, though not everyone concurs with the perceptions of those who named them (Figure 3.4).

The *malleus*, the first of the three ossicles, is in contact with the tympanic membrane. Like a mallet, it has a head, neck, and handle. The handle is directly attached to the connective tissue of the tympanic membrane at the center of the drum. It is this central attachment which causes the eardrum to assume its characteristic conical shape. The inner surface of the head of the malleus is sculptured to provide a snug fit with the second of the two ossicles, the anvil or incus.

The *incus* is shaped somewhat like an anvil, consisting of a broad body from which two processes project. The anterior surface is hollowed out to accommodate the shaped surface of the malleus in an articulatory link. The long process projects downward from the body of the incus and at the extreme end turns sharply inward (medially), where it ends as a rounded, button-like projection called the *lenticular process*. This process provides the articulatory contact with the last of the three ossicles, the stapes.

The *stapes*, or stirrup, has a head shaped so as to receive the lenticular process of the incus. The neck branches into the two arches (crura) that connect the neck to the footplate. The footplate of the stapes, composed of bone and cartilage, is the medial end of the ossicular chain. It is inserted into the oval window in the cavity wall separating the middle ear from the inner ear. Thus it represents the interface, or common boundary, between two systems: the mechanical system of the tympanic membrane and ossicular chain, and the hydraulic system of the fluid-filled inner ear on the other side of the oval and round windows.

The ossicular chain stretches, then, from the tympanic membrane to

Figure 3.4 The human ossicles of the middle ear reproduced in modified form.

the oval window. It is not a true suspension bridge, however, for it is not suspended exclusively from these two end points, but additionally by five ligaments; three support the malleus, running from different points to the walls of the tympanic cavity, one supports the incus, while the fifth (the *annular*, or ringshaped, ligament) holds the stapes in the oval window. This annular ligament is inserted into the bony walls of the oval window and is attached to a thin cartilage covering the footplate of the stapes.

These ligaments serve to suspend the ossicular chain in such a way as to achieve a balanced system which offers little inertia.* This low inertia makes the system very sensitive to movements of the ear drum and also provides for a rapid damping of ossicular vibration once the sound energy ceases. This is an important factor, since if the vibrations were to continue, even for a brief period, they would result in the addition of wave form components foreign to the original pattern. Such alien vibrations, termed distortion, increase the difficulty of the perceptual task of restructuring the original message.

The middle ear mechanism is acted upon by two muscles, the *tensor tympani* and the *stapedius*. The action of these muscles is believed to relate to the reduction of sensitivity of the chain to high-intensity sound waves, a protective mechanism we shall not include in our discussion.

To recapitulate, we have seen how sound energy travels from the outer ear to the interface between the middle and inner ear at the oval window. We also noted that the middle ear mechanism is necessary to match the impedances of two systems lacking a high degree of compatibility for energy transfer. Low compatibility really signifies a difference in density and elasticity between two systems.**

In the ear compatibility between the gaseous medium through which sound waves travel and the fluid medium of the inner ear is low. The loss of energy resulting from this mismatch would be in excess of 30 dB (Wever and Lawrence, 1954). Were it not for the impedance-matching function of the eardrum and the ossicular chain, it would be necessary for the speaker to shout in order for us to hear normal conversation at the accustomed loudness level.

If it were possible to stand on one of the crura of the stapes and look

---

*Inertia is the tendency of matter to remain at rest, or if moving, to keep moving in the same direction.
**In a text on physiological acoustics. Wever and Lawrence (1954) have shown that the impedance mismatch between air and water is so great that if one shouts to a swimmer who is under water in a swimming pool, 99.9 percent of the sound energy will be reflected from the surface into the air. Only one-tenth of one percent will enter the medium of the water. Not a very effective means of communication!

back across the ossicular chain, it would be apparent that the area of the tympanic membrane is considerably greater than the area of the footplate of the stapes behind you. The conductive mechanism of the middle ear thus effects a funneling action on the sound energy. Energy collected over the relatively large area of the drum membrane is concentrated by the ossicular chain on the much smaller area (about fourteen times smaller) of the footplate of the stapes. There is, in addition, approximately a further 3 dB gain which results from the ossicular chain functioning as a complex lever system. Zemlin (1968, p. 395) and Harris (1974a, p. 32) have computed the total mechanical advantage of the middle ear mechanism to be about 25 to 27 dB. Thus our bridge serves not only as a conductor of sound across the cavity of the middle ear, but also as an effective transformer providing an efficient impedance-matching function.

The sound wave energy has now been conveyed to the oval window with minimal energy loss and without distortion of the distribution of energy across the various component frequencies. The all-important pattern remains intact.

The Inner Ear

However impressed we may be with the evolutionary success in solving the problem of sound conduction to the oval window, it remains a mechanical achievement. The next problem is far more complex, for it involves getting the patterned information into the nervous system of the organism. To do this it was first necessary to solve the problem of conveying the energy from the oval window to the sensory organ of hearing, the *organ of Corti*, which constitutes the interface between the external and internal components of the auditory system.

To better understand how this is achieved, let us examine the architecture of the inner ear.

The Cochlea.   On the internal side of the oval window of the right ear lies a complex and extensive structure which is buried in the temporal bone of the skull (Figure 3.5). This comprises the *cochlea* which houses the organ of hearing, and the *semicircular canals*, which house part of the balance-regulating system. Immediately behind the window is a large egg-shaped cavity called the *vestibule*. To the left are what appear to be five alcoves with passages leading away from them (in the left ear everything is reversed). These are the fluid-filled semicircular canals of the *vestibular apparatus* of the ear, which constitute an important part of the balance

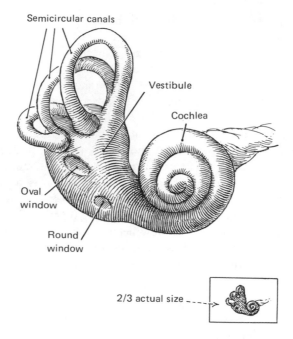

Figure 3.5 The bony labyrinth of the inner ear partly open
to show the membranous labyrinth within it.

system of the body. Although this apparatus is anatomically related to the
structure of the sensory end-organ of hearing, their relationship in the
perceptual process is slight and indirect in nature. For this reason, we will
proceed to consider another channel to the right.

Exploration of this channel, which is also fluid-filled, reveals two
interesting facts. The first is that the channel spirals upward for 2¾ turns.
Because of its resemblance to a spiral staircase, this channel has been
named the *scala vestibuli* (vestibular staircase). At the top of the spiral
(the apex) is a small opening termed the *helicotrema*; this leads to a similar channel, or scala, which spirals the same distance downward. The
second interesting discovery is that this scala terminates at a round membrane, the *round window*. This was seen from the other side of the bony
wall when we surveyed the middle ear or tympanic cavity. This descending scala is known as the *scala tympani.*

We have identified two windows, the oval and round windows, each
communicating with one of the scalae. We have also seen that these two
scalae, the scala vestibuli which begins at the oval window and the scala
tympani which terminates at the round window, are a continuous channel,
since they join at the helicotrema.

Let us now turn our attention to the physical structure of the organ of hearing. The spiral bony structure (the otic capsule) is shaped like a coiled shell, thus its name, the *cochlea* (Figure 3.6). The floor of the scala vestibuli and the roof of the scala tympani both consist of membranes. They run from a narrow bony shelf (*spiral lamina*), which projects a short distance into the cochlear space, to the wall opposite where they are attached by a ligament (*spiral ligament*)

The membrane of the scala vestibuli, known as the *vestibular membrane* or alternatively as *Reissner's membrane*, runs at an angle of about 45° from the spiral shelf, while the membrane which forms the roof of the scala tympani projects straight across the cochlear cavity. The angle of these two membranes thus creates a third middle scala, the *scala media* (also known as the *cochlear duct*). Since the membrane separating the scala media from the scala tympani provides a base for the sensory end-organ of hearing, the organ of Corti and its related structures, the dividing membrane is termed the base or *basilar membrane*.

We thus have three scalae. However, unlike the scalae vestibuli and tympani, the scala media is a completely self-contained space with no connection to either of the other scalae or to the vestibule. The only connection it has is to the semicircular canals by way of a small duct, but there is dispute over whether this remains open in adults. The fluid (endolymph) within the scala media differs in chemical composition from the perilymph of the other two scalae.

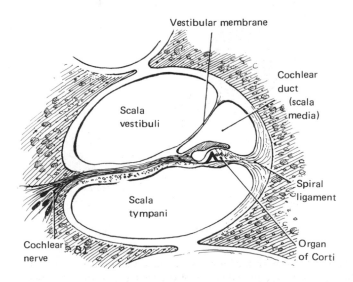

Figure 3.6 A diagram of a cross section of the cochlea.

Before we consider the sensory organ of Corti, which constitutes the transducer of the hearing apparatus, it is both necessary and timely to relate this anatomy to that function which is our main concern: the ability of the nervous system to replicate the information patterns of the acoustic signal into patterns of neural discharges which result in perception.

We recall that our energy wave had progressed in mechanical form to the footplate of the oval window; now it must enter into the inner ear. The perilymph and also, as we shall see, the endolymph, provide the next medium for energy transmission, a hydraulic medium. One of the properties of fluids is that they are practically incompressible, which accounts for the fact that we can row a boat in water but not in air. If we push a cork into the narrow neck of a full bottle, the liquid will squirt its way out between the cork and the glass. The fluids of the inner ear are enclosed within the nonelastic bony walls of the cochlea and vestibular apparatus. The movement of the stapes against the noncompressible fluid would be virtually impossible without a means of releasing the increased and decreased pressures of the energy wave. This element of "give" in the system is provided for by the continuity of the channel from the scala vestibuli via the helicotrema to the round window at the end of the scala tympani. Figure 3.7 illustrates schematically the movement of the stapes in the oval window and the reciprocal movement of the membrane of the round window which permits energy to flow through the fluid medium.

This mutual relationship between the windows allows for the displacement of the fluids necessary for the energy flow. The movement of

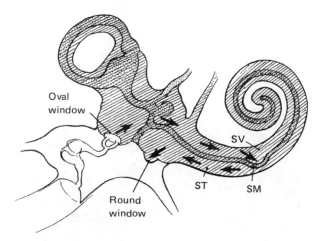

Figure 3.7 The path of a pressure wave through the cochlea, illustrating the reciprocal action of the oval and round windows. (ST-Scala tympani, SM-Scala media, SV-Scala vestibuli)

the stapes about a neutral position of rest will replicate the positive and negative pressures which the sound wave impresses on the tympanic membrane and will in turn effect the same pattern of movement in the perilymphatic fluid. Thus the essential pattern of information has been passed into the inner ear, but has not yet entered the nervous system. To do this the sensory nerve endings of hearing must be stimulated; these are integral components of the *organ of Corti*, which stands on the basilar membrane, bathed in the endolymphatic fluid. Let us first examine the organ of Corti; then we will consider how the energy-pressure wave is transmitted, almost instantaneously, down the cochlear duct.

The Organ of Corti.   We know now that the basilar membrane separates the scala media (cochlear duct) from the scala tympani. It extends from the bony plate (spiral lamina) to the opposite wall to which it is attached by the spiral ligament. We know also that above it, rising at a 45° angle, is Reissner's membrane, which separates the scala media from the scala vestibuli (Figure 3.8). The organ of Corti in the scala media is therefore in a self-contained space filled with endolymph. Now look at Figure 3.9, which details this critical energy-converting unit. To this point our energy pattern has been mechanically passed from one unit of the system to the next. Even the transfer from the ossicular chain to the hydraulic system of the inner ear fluids was a fairly simple event. The change which occurs at the interface between the endolymph and the sensory end-organ is far more complex, for not only does transduction of energy form occur, but the first stage of pattern analysis takes place. The change which occurs at the interface between the endolymph and the sensory cells is quite complex, for the organ of Corti has mechanical, electrical, and probably chemical functions.

Supportive Cells and Structures.   The organ of Corti is named after the man who first described it in 1836. It is made up of (1) the sensory receptor cells and a network of cells and cellular structures that supply them with both architectural support and nutrition; (2) a delicate membrane (the tectorial membrane) which plays a critical role in stimulating the hair cells; and finally (3) a nerve-fiber system which runs both to and from the sensory cells.

The most obvious components of the structural supports are the two central pillars or *rods of Corti*. These are comprised of supportive cells and stand on the basilar membrane with widespread bases. The two rods come together at the top forming a teardrop-shaped tunnel. Close support for the sensory cells is also provided by tall columnar structures known as *phalanges*, each consisting of a single cell having a nucleus near its

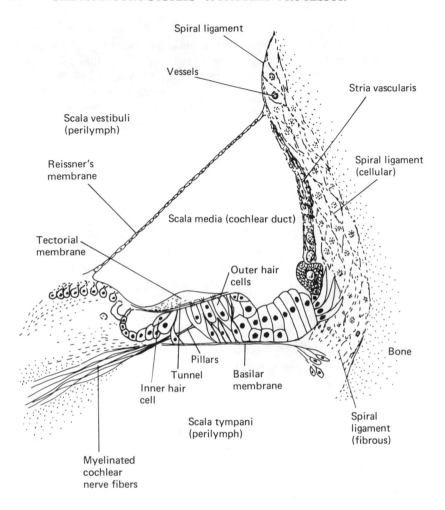

Figure 3.8   Diagram of a section of the cochlear duct.

base. These stand on the basilar membrane, or for some (inner phalanges), on the spiral lamina. The apices of these various supporting cells, several of which for simplicity are not labeled in the diagram or discussed in the text, join together to form a thin membranous net which lies across them, providing further support for the tops of the sensory cells.

Before we examine the sensory cells themselves, there remains one important component to describe, a very delicate diaphanous structure: the *tectorial membrane*. This membrane is attached to the projection of the spiral lamina at the inner end and extends over the tips of the sensory cells and merges with the supportive cells at the far side of the organ of

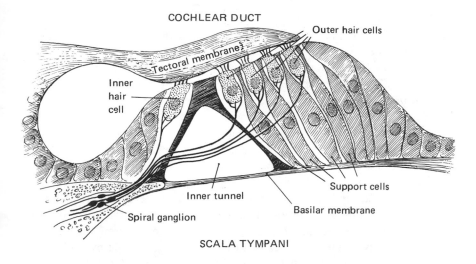

Figure 3.9 A close-up diagram of the organ of Corti.

Corti. Because of its extremely flimsy composition, it is capable of being wafted by any currents which flow through the endolymphatic fluid around it, like a delicate scarf held in the breeze. It is this membrane which serves to pass to the sensory cells the movement pattern impressed initially upon the fluids at the oval window.

The Sensory Receptor Cells. The final interface between the auditory pathway of the nervous system and the mechanical-hydraulic system which conveys the energy pattern from the outer ear occurs at the sensory end organs of hearing. I mentioned earlier that these cells represent three functions: a mechanical aspect, an electrical aspect, and possibly a chemical aspect. It is here that the acoustic energy wave, which has existed in the various media of air, membranes, bone, and fluid as variations in pressure is transformed into a totally different chemical and neural form.

There are two groups of sensory receptors totalling approximately 12,000 to 15,000 cells. On the inner side of the pillars of Corti are a single row of teardrop-shaped hair cells; on the outer side there are three to five rows of cylindrical-shaped outer cells, with the greater number toward the apex. Each hair cell, inner and outer, is topped by rows of small cilia (hairs) which give them the name *hair cells*. The number of cilia varies from 40 to 60 on the inner cells, increasing in number as the cells are located closer to the apex of the cochlea. The hairs project above the membranous net which connects and supports the tops of the supporting cells. They are in some way lightly connected to the tectorial membrane, though how this is achieved is still not understood (Zemlin, 1968, p. 410). However, direct contact between the hairs and the tectorial membrane is essential to the transducer function of the organ of Corti.

The mechanical action of the organ of Corti is now fairly well understood. This has been described by Békésy (1960) in his book *Experiments in Hearing*. We shall very soon be considering the complete energy flow from the eardrum to the hair cells; at this point it is sufficient to say that pressure waves passing through the endolymph of the scala media cause a downward (positive) or upward (negative) deflection of the basilar membrane and the tectorial membrane. However, the angle of displacement of these two membranes differs because, while the basilar membrane is attached at both ends and is caused to stretch, the tectorial membrane is hinged only at one end, the other end being loosely attached to the supporting cells of the organ of Corti. This difference between the angle of incline of the two membranes causes them, in effect, to slide by each other. This results in a sideways deflection (shearing action), of the cilia toward the inner wall for a downward deflection (positive pressure) and toward the outer wall for an upward deflection (negative pressure). Since the hair cells stand on the basilar membrane with their cilia attached to the tectorial membrane, this shearing action will result in the hairs being bent to accommodate the difference in the displacement of the two membranes, as in Figure 3.10 (Whitfield, 1967, p. 27–28).

Energy Transmission Through the Cochlea.   Let us pause now to place what we have discussed within the larger perspective of the transmission of the energy pattern. Indeed, it might be helpful to review the process of the wave pattern from the air to the hair cell.

The complex pattern of vibration is impressed upon the air medium by the vibration source. The pattern exists in terms of an enormous range of small positive and negative pressure variations around normal atmospheric pressure. These variations of pressure travel in waves through the air like patterns of ripples across a pond. As these sound waves enter the external ear canal they cause the eardrum to be depressed inward from its rest position when the pressure is greater than normal air pressure, and to be distended outward when there is a negative or lowered pressure. The movement of the tympanic membrane thus replicates the pattern of positive and negative variations in air pressure, retaining the integrity of the information-bearing sound wave. This pattern of drum movement sets in vibration the ossicular chain, which functions as a conducting unit passing the energy across the middle ear cavity to the oval window in the bony wall dividing the middle and inner ear. The chain also concentrates the energy gathered over the relatively large area of the tympanic membrane

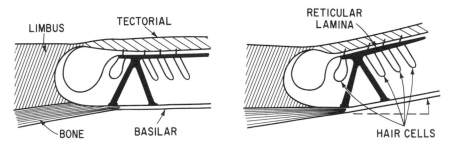

Figure 3.10 Movement of the organ of Corti and the tectorial membrane. The shearing action between these two stiff structures bends or shears the hairs of the hair cells. (Davis, Chap. 2, 1960)

onto the much smaller area of the footplate of the stapes. The patterned movement of the stapes sets up pressure waves in the perilymphatic fluid which fills the vestibule on the other side of the oval window. These pressure waves travel into and along the scala vestibuli.

At this point we must discuss the dual pathway these energy waves can travel. We have already seen that energy can flow through the perilymph to the apex of the cochlea where the sound waves may pass from the scala vestibuli through the helicotrema to the scala tympani, then down to the end of the spiral to be released through the round window. However, it is the second pathway which leads to the stimulation of the hair cells.

As we recall from the previous chapter, one of the measures of the sound wave is the frequency at which it vibrates. A measure related to frequency is the length of the wave. *The wavelength is inversely proportional to the frequency.* Since the measurements are made over a period of one second, if we double the frequency the wavelength must decrease in order to achieve a greater number of vibrations within the one second period (Figure 3.11). This fact is important to an understanding of the manner of stimulation of the organ of Corti.

Consider again the sound energy entering the cochlea either via the stapes and oval window (air conduction) or via the bony walls of the otic capsule (cochlea). This energy causes pressure waves to spread almost instantaneously throughout the cochlea. The pressure wave causes the basilar membrane to respond mechanically, a much slower response than the hydraulic vibrations. Because of the special geometric and elastic properties of the basilar membrane, the energy vibration always travels from the stiff, narrow, basal portion toward the more flaccid, wider, apical portion of the basilar membrane. This happens regardless of whether the sound enters the cochlea through the oval window, the walls of the otic

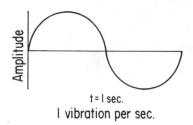

Figure 3.11   Wavelength must decrease as the number of vibrations per second (frequency)

capsule, or through an artificial stapes placed in the apical turn (Wever and Lawrence, 1954, pp. 271–272). The mechanical response of the basilar membrane to the pressure wave is in the form of a traveling wave which, as previously stated, begins at the basal end and travels toward the apex, increasing in amplitude until it reaches a particular point, depending upon its wavelength and, therefore, its frequency. The amplitude of the wave then falls off sharply to an approximately constant displacement (Rhode, 1971). The point of maximum displacement along the basilar membrane is a continuous function of the frequency of the stimulus, with the higher frequencies stimulating the basal region and the lower frequencies stimulating the more apical portions. The velocity of the traveling wave is essentially constant and independent of frequency up to the point of maximum displacement, but increases beyond that point.*

The relative movement of the tectorial and basilar membranes exerts a shearing force on the cilia of the hair cells, bending them mechanically. The energy then passes through the basilar membrane into the perilymph of the scala tympani and is vented through the round window.

Stimulation of the Hair Cells.   The way in which the mechanical movement of the hair cells induces an electrical impulse in the nerve fiber

---

*I am indebted for this description of cochlear hydrodynamics to my colleague Berner Chesnutt, University of Texas Medical Branch, Galveston, Texas.

running from the hair cells to the auditory nerve has been subject to considerable investigation (Davis, 1960, 1962); however, the exact process remains to be determined (Harris, 1974b). It is known that the neurophysiology of the cochlea is a highly complex process related to the neurophysiology of the body as a whole. Research has shown that the chemical composition of living cells provides for the conduction of electricity. It is known that the potassium and sodium ions of the cell in its resting state produce an electrical balance between the cellular cytoplasm and the outside of the membranous wall which surrounds the cell. If the cell is irritated in some way, the state of rest is disturbed and the electrical balance upset, resulting in a flow of current along its attached nerve fiber. This latent capacity to induce electrical activity is known as *electrical potential*.

The hair cells of the cochlea possess such a potential. A direct relationship is known to exist between the pattern of mechanical vibration of the basilar membrane on which the hair cells stand and the associated pattern of nerve impulses in the auditory nerve. The hair cells in some way permit the internalization of the mechanical pattern. It is believed that the shearing action on the cilia disturbs the hair cell membrane which in turn upsets the electrical balance by changing the electrical potential. It has been suggested that this electrical potential may cause the release of a chemical transmitter (Davis, 1960), or that the mechanical disturbance itself causes transmitter release. The chemical transmitter depolarizes the nerve endings which form chemical synapses with the hair cell (Whitfield, 1967, p. 47–50).

There are other complicated electrical relationships (potentials) existing between various parts of the cochlea. The most notable evidence of these are *cochlear microphonics*—small amounts of electrical energy which can be measured within the cochlea. These electrical charges are apparently generated by the hair cells and are capable of reproducing both the frequency and amplitude of a sound stimulus; however, there is at present no evidence to suggest that they are important to auditory perception, so we shall not consider them further.

Theories of Hearing—Frequency

So far our discussion of the cochlea as a transducer has considered only a single cell. It is obvious that the information in the energy wave cannot be replicated in the auditory pathway by one cell; in other words, a single cell is incapable of internalizing the total speech wave. We know that the traveling wave pattern is complex, resulting from the interaction of multiple vibrations. The question is how this complexity can be internalized.

For many years a theory proposed by a German physiologist, H. von Helmholtz (1863), dominated the understanding of this process. He postulated that the ear functioned as a series of tuned resonators, each of which triggered a neural impulse whenever the input sound wave contained frequency components which coincided with its resonant frequency. This concept can be demonstrated by placing small pieces of tissue paper between a number of strings on the piano. If you then sing a loud note close to them some will be seen to vibrate: Components of your vocal sound coincide with some of the frequencies of the piano strings and set up sympathetic vibrations in strings at or around those frequencies. Helmholtz first suggested that the rods of Corti served as the tuned resonators, but it was soon found that there are more discriminable tones than there are rods. He then suggested that it was the basilar membrane which served as a tuned resonator. Subsequent experimentation and direct observation have shown that the basilar membrane when stimulated, even by simple tones (pure tones), vibrates not at one point, as a tuned resonator would, but along much of its length. Furthermore, its patterns of vibration overlap for similar tones.

Resonance Theory.   Helmholtz's *resonance theory*, which is dependent upon place of stimulation, is known, therefore, to have been an invalid explanation of how the organ of Corti responds to the mechanical energy wave.

The theory of hearing proposed by Helmholtz suggested that specific hair cells and their associated nerve fibers at specific locations on the basilar membrane were stimulated by a resonant response to discrete components of the complex wave, thus providing a one-frequency, one-nerve fiber cochlear analysis of the waveform. This *resonance theory* had to be rejected because of the evidence that the organ of Corti responds, not by discrete cells and fibers, but by patterns of activity generated by groups of cells.

Frequency Theory.   In 1886 Rutherford suggested that the frequency of a component wave might be transmitted in terms of the number of pulses per second in a nerve fiber. This *frequency theory* directly equated the frequency-component information in the mechanical energy wave with the rate of firing of the nerve impulses. Reproduction of the complexity of the acoustic pattern was dependent upon the pattern of fibers involved: a dependence on a place mechanism. The frequency theory in its original form had to be rejected when it was subsequently demonstrated that fibers cannot fire at a rate much greater than 300 impulses per second, whereas the human ear responds to frequency vibrations up to 20,000 Hz. Certain mammals respond to much higher frequencies.

Volley Theory. In 1949 Wever modified Rutherford's theory to overcome this limitation. He suggested that groups of nerve fibers fire in a synchronized manner. Acording to this theory, five fibers, for example, would fire a volley of impulses one after the other, providing a combined impulse rate five times greater than the maximum rate of a single fiber. The volley principle allows for additional nerve fibers to cooperate in the internalization of the frequency information as the maximum rate of a fiber(s) is exceeded.

Like all theories to date, the volley theory does not completely explain the internalization of frequency information. It depends upon the synchrony of multiple nerve impulses, but the accuracy of the synchrony begins to break down around 1000 Hz and is lost by about 3000 Hz. (Whitfield, 1967, p. 145). It is, however, a valid theory for the accurate processing of frequency information up to about 500 Hz and provides less precise data up to 3000 Hz.

We have not yet discussed a theory which will explain the processing of stimuli from approximately 3000 to 20,000 Hz. For this we must turn to one which, ironically, like Helmholtz's resonance theory, is dependent upon place of stimulation, but, unlike his theory, does not involve the concept of resonance. The place theory, in the form explained by G. von Békésy, depends upon a knowledge of the patterns of vibration of the basilar membrane.

Traveling Wave Theory. As was explained earlier, when the stapes moves in the oval window, pressure is propagated almost instantaneously throughout the cochlea, though the traveling wave on the basilar membrane develops somewhat more slowly (Békésy, 1947). The wave travels from the relatively stiff basal end of the basilar membrane to the less stiff region of the apex. The presence of such *traveling waves* in the cochlea has been demonstrated by Békésy who observed them when illuminating the vibrating membrane with a stroboscopic light, a flashing light which is capable of making moving objects appear stationary. The peak of the wave was seen to occur at different points along the membrane, the locus depending upon the frequency of the stimulus. The amplitude of the wave increases as it travels along the membrane toward the point at which its peak amplitude is reached, after which it falls away steeply to a very small displacement (Békésy, 1960; Rhode, 1971). The displacement of the membrane at its maximal point appears to create small swirling currents (*eddies*) in the endolymphatic fluid for intense stimulation. These eddies follow the position of the maximum vibration. Békésy suggested that these fluid disturbances may be responsible for exerting a concentration of pressure on the hair cells at that point. This would provide for a narrowing of the actual stimulus area even when the membrane is vibrating along most of its length.

Each frequency has been observed to result in a pattern of membrane movement unique to that frequency. This arises from the discreteness of the locus of the maximum amplitude and the extent of the movement, which is dependent upon the intensity of the stimulating sound wave. Békésy's theory depends, therefore, upon the relationship between the energy pattern of the original stimulus and the equivalent pattern of vibratory movement of the basilar membrane. It has been pointed out that the effect of the vibration is further concentrated by the creation of eddies in the fluid. These are set up at the point of maximum displacement of the membrane providing for a fairly specific localized response of hair cells at that location.

Place-Volley Theory.   Current opinion is that no single theory of hearing proposed to date can completely account for how the frequency information is transformed from patterns of fluid movement to patterns of neural impulses. It is generally accepted that a combination of a place theory such as the one proposed by Békésy, covering frequencies from around 3000 Hz up, and a volley theory such as Wever's for frequencies up to 500 Hz, with an overlap of the two theories in the mid-frequency range, is necessary to explain the internalization of information represented by energy at both high and low frequencies. There is, of course, no reason to assume that the organ of Corti behaves in one manner for all frequencies. It might prove essential to develop two explanations for two modes of behavior. Whitfield (1967, p. 146) is not satisfied with this place-volley theory of hearing. He emphasizes that pitch recognition cannot be a peripheral function in the sense that each discriminable tone is represented by a single fiber, or even by groups of fibers. He acknowledges Békésy's evidence that the basilar membrane acts as a tapered, low-frequency pass filter, resulting in the point of maximal stimulation of the membrane moving from the basal end toward the apex as the frequency of the stimulating tone is lowered. At the same time, he points out the weakness of the place theory: namely that any disturbance results in the movement of a considerable part of the membrane.

Whitfield challenges the volley theory because, he claims, it is not only untenable for frequencies above 1000 Hz where the synchrony of firing begins to break down, but also because he considers the hypothesis explaining low tones to be weak. The problem of:

> . . . detecting the time differences between two successive pulses in the *same* channel, and comparing this with the slightly different interval arising from a later presentation of a second tone (the usual

frequency discrimination situation) . . . requires a mechanism for which no physiological evidence has yet been uncovered. . . . (1967, p. 146).

The general weight of evidence then indicates that, though information about the stimulus frequency is available over part of the frequency range in the form of the intervals between nerve impulses, this is an epiphenomenon which is not made use of by the nervous system for stimulus frequency identity. (ibid.)

Whitfield's Boundary Theory.   As an alternative to the place-volley theory, Whitfield proposes a theory based upon recognition by the auditory system of the *pattern of fibers* activated at the periphery. He points out that:

Each frequency gives rise to a unique configuration, since both the extent of the activity and the position of the amplitude maximum change with frequency. Each pattern contains, therefore, the necessary data to define the frequency and intensity (1967, p. 147).

The fact that a stimulus tone activates a wide array of fibers on either side of the stimulus frequency means that tones close in frequency share a large number of overlapping fibers. If the intensity of a tone is increased, the boundaries of the area stimulated are extended, though the relative magnitude of the response decreases as one moves away from the point of maximum stimulation. These boundary characteristics are transformed and emphasized in the cochlear nucleus, the first neural processing center of the auditory pathway (Figure 3.12). Whitfield maintains that the stimulus frequency is signaled by the position of these boundaries between the active and inactive fibers.

The effect of the transformation we have been discussing is to produce, within the fiber array, regions of activity bounded by regions of inactivity. In this transformation components of different frequency are represented by the different blocks of active fibers, while the intensity of each is represented by the number of active fibers ("width") in the active block. This pattern can of course change with time in the manner shown. Such a coding arrangement would seem well suited to the properties of the nervous system. The simple requirement of activity or nonactivity in the pathway makes the representation very resistant to degradation by spontaneous "noise" impulses, or to differences in the exact temporal distribution of pulses which occur when successive identical stimuli are given. Fur-

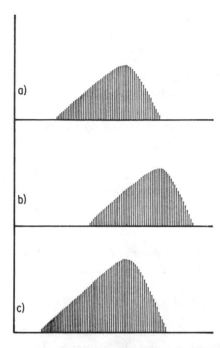

Figure 3.12 The distribution of activity in the array of auditory nerve fibres for two different stimulus frequencies, (a) (b). Each vertical bar represents a small group of adjacent fibers, and its height represents the the mean discharge rate of those fibers. (c) represents the response to a stimulus of the same frequency as (a) but of higher intensity. The response involves both higher discharge rates and more fibres. (Whitfield, 1967)

thermore, the nervous system appears much better adapted to detecting the boundaries between regions of activity and inactivity than it does to locating the positions of maximum and minimum displacement. (1967, p. 153)

Whitfield's approach to explaining the manner in which auditory information is initially processed is very compatible with the concept of auditory perception as a process based upon pattern recognition.

Theories of Hearing—Intensity

As we learned, the frequency structure of the sound wave is only one of the three measures of its pattern; intensity of individual components will, to varying degrees for different sounds, be critical, or at least important, to auditory perception. The intensity of the components of the complex sound wave varies over time. It has been pointed out that frequency and intensity in the complex pattern are not discrete elements but two measures of one event. If we increase the amount of energy at any section of the range of frequencies contained within the complex wave, the overall intensity will change, but so also will the overall pattern of vibration. When we break down a complex tone into its component frequencies, we are in fact determining not only which frequencies it contains but also how much energy is present at each of those frequencies, or within certain frequency bands. It is the interaction of these two measures which, to a major extent, determines the pattern.

We have seen that the internalization of frequency information can be considered primarily dependent upon boundary patterns of active fibers at points along the basilar membrane. Intensity, on the other hand, appears to depend upon the *number* of fibers activated. It is known that the stimulus energy necessary to trigger a nerve impulse is different for different fibers, and that the rate at which fibers fire increases with stimulus energy. Whitfield (1967, p. 164) has inferred from these observations, that the hair cells themselves have different sensitivities to mechanical stimulation. He considers the auditory nerve fiber threshold for intensity to be determined by whether it originates from an inner hair cell, which appears to have a relatively high threshold, or an outer hair cell, with lower thresholds, and also by the number of cells involved.

> The proposal that the auditory threshold of a nerve fiber depends not only on whether it takes origin from an inner or an outer hair cell but also on the number of hair cells innervated, means that it is not necessary to postulate two distinct populations of fibers with respect to threshold. A suitable combination of the two mechanisms can result, in the limit, in an almost continuous distribution of threshold as is found experimentally (1967, p. 167).

There remains to be considered in this chapter what lies between the transformation of the energy wave into a neural impulse pattern at the organ of Corti, and the processing of the speech sound by the auditory cortex. Before we consider this, however, a review of the neurophysiological behavior of the organ of Corti might be helpful for the reader who may feel a little overcome by the details.

Review of Inner Ear Functions

We began our discussion of the manner in which the organ of Corti transforms hydromechanical energy in the endolymph into electrical impulses in the cochlear nerve fibers by discussing the structure of the organ. We saw how the sensory end organ hair cells (ciliated cells), which are critical to the transformer process, stand on the basilar membrane supported by a network of other cells, which also provide nutrition. The hair cells were described as constituting two groups (inner and outer hair cells), totaling approximately 20,000 cells extending along the entire length of the basilar membrane, but concentrated at the apex of the cochlea. The cilia, or hairs, of the cells are connected to the tectorial membrane which floats above them. This connection, between the hair cells standing on the basilar membrane and the tectorial membrane above them, allows for a shearing action to occur when the hydraulic pressure wave, set up initially by the movement of the stapes in the oval window, causes a differential displacement between the two membranes. We saw that the firing of the neural impulse resulted from the disruption of the electrical balance of the hair cell by the mechanical movement of the cilia, somehow triggering an electrical impulse in the fibers associated with it.

We continued our examination of how the cells internalize the information necessary for pattern reconstruction by considering several theories of hearing. We saw that neither the place theory nor the frequency theory alone could explain how frequency and intensity information is internalized. We then ·examined a theory employing both the concept of synchronous nerve firings (frequency volley) to account for low frequency detection and a place concept (place of maximal displacement) for component vibrations over 3000 Hz, with an overlap of functions in the midfrequency range. Although this is at present the most widely accepted theory, we also reviewed the work of two other researchers: Békésy, who proposes a traveling wave theory, that is, a modified place theory, and Whitfield, who has elaborated upon Békésy's evidence to develop a theory based upon identification of boundaries of stimulation. Finally we asked ourselves how intensity information might be internalized and concluded that it probably results from the number of fibers triggered and from the differential sensitivity of fibers.

The Auditory Pathways

The Afferent System. Our interest in the neurological pathways, which run from the organ of Corti to the cortex of the brain, does not stem from a need to understand anatomy per se but rather to fathom the role they play in facilitating the perception of speech sounds. In Chapter 5 we shall consider this process in detail. At that time we shall hypothesize the possible manner in which the neuroanatomy of the system serves to determine the nature of the perception. Meanwhile, we will trace the neural pathways of the auditory system (Figure 3.13).

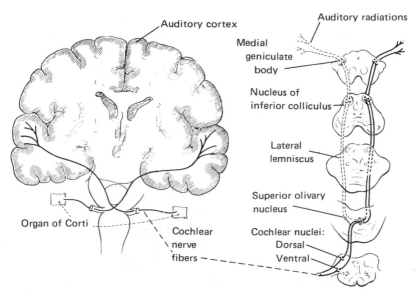

Figure 3.13   A diagram of the conscious auditory pathway.

All of the inner hair cells on the basilar membrane are served by several nerve fibers, and each fiber runs to several cells. The outer and inner hair cells are served by different sets of fibers, with some overlap between them. The significance of this distribution has not yet been clearly determined, though it may have something to do with the analysis of intensity information.

Each fiber originates from a cell body in the long *spiral ganglion*, a mass of cells which runs through a canal in the central bony section of the cochlea, following the spiral turns. At each level, peripheral fibers leave the ganglion to serve the organ of Corti.

Traveling in the opposite direction, that is, away from the periphery and toward the cortex, fibers leave the spiral ganglion, passing to the center of the cochlea where they come together to form the acoustic nerve. The fibers from the lower turns of the cochlea twist around the straight ones running directly from the apex. Altogether, about 30,000 individual neurons are involved.

After the acoustic nerve fibers leave the central core of the cochlea, they enter the bony tunnel at its base, the *internal auditory meatus*, where they are joined by the vestibular nerve which serves the semicircular canals of the vestibular apparatus. These two nerves together form the *auditory nerve* (VIIIth cranial nerve), which very soon enters the midbrain at the level of the medulla and pons. As soon as it does, the auditory fibers branch off to the first of a series of neural centers, the *cochlear nucleus*.

In the cochlear nucleus the fibers divide into two branches, one branch traveling to the dorsal portion (*dorsal cochlear nucleus*) and the other to the ventral portion (*ventral cochlear nucleus*).

This mass of cell bodies constitutes a kind of substation of the auditory pathway, linking the fibers from the spiral ganglion to second level (*second order*) fibers that will continue upward to higher neural centers. The fibers are known to follow two routes, a direct and an indirect, via various way-stations or processing centers in the lower brain stem.

More than half the fibers cross (decussate) to the opposite side of the medulla by way of the *trapezoid body*. Some fibers travel upward by the direct route to the midbrain, where they terminate in a neural center known as the *inferior colliculus*. Other fibers, crossing from the cochlear nucleus, terminate either in the *superior olivary complex* or in the *nucleus of the lateral lemniscus*, centers from which other neurons leave to travel upward. None of the fibers which remain on the same side of the cochlear nucleus from which they originate, follows a direct route to the inferior colliculus. They all travel by way of the superior olivary complex either directly to the inferior colliculus or by way of the nucleus of the lateral lemniscus. At the level of the inferior colliculus, some fibers cross back to the side of origin, while other fibers travel from the uncrossed route to the crossed route; thus, decussation occurs at two levels in the auditory pathway.

Fibers then leave the inferior colliculus to travel to the *medial geniculate body* from where fourth-order neurons extend to the auditory cortex.

The following points are made by way of clarification:

The pathway we have examined for one ear is mirrored on the other side. Thus the bulk of fibers originating at the cochlear nucleus for each ear cross to the side of the opposite (contralateral) ear. Both ears send a portion of the fibers upward on the same (ipsilateral) side. As a result of this dual pathway, fibers from each cochlear nucleus travel both to the auditory cortex of the same hemisphere and to the cortex of the opposite hemisphere, providing each side of the brain with information from both ears. This bilateral innervation provides a back-up system for the auditory cortex, preventing total deafness in one ear that would otherwise result from destruction of the neural pathway on the opposite side. It also provides each cortex with two sets of data, an important factor for certain auditory skills.

We have talked about fibers running upward from the cochlear nucleus to the auditory cortex. It is important to make clear that we are concerned not with a single set of fibers, but with a relay system of fibers with interconnections in the various brain stem nuclei. An analogy might be made to a well organized bus system. Direct, nonstop service is provided from each cochlear nucleus to the inferior colliculus of the opposite side. Similar nonstop service is provided from the superior olivary complex to the inferior colliculus, both for crossed and uncrossed fibers. Thus a fiber synapsing with a cell in the inferior colliculus may have arrived directly from the cochlear nucleus or may have arrived from the superior olivary complex. Local service is provided between each of the nuclei, but to travel by this route it is necessary to change buses, or in this case, fibers. Therefore, the fibers which finally reach the auditory cortex in the temporal lobe of the brain did not originate at the cochlear nucleus, but are *at least* fourth-order neurons.

The Efferent System.   The pathways we have been considering are all afferent (sensory), or ascending, pathways leading from each nucleus upward to higher centers of the brain and, ultimately, to the auditory cortex. The concept of the senses as passive systems conveying sensations to the cortex for analysis and evaluation was at one time in accord with what was known about the neuroanatomy of the sensory pathways. Despite the appearance around the turn of the century of evidence (Lorente de Nó, 1933) which showed the presence of fibers running downward in the auditory pathway (efferent fibers) in the opposite direction to the sensory fibers, there was little or no follow-up until Rasmussen's (1942, 1946) early work of the 1940s. Rasmussen's subsequent research (1953, 1955, 1958, 1960) has conclusively demonstrated the existence of a system of efferent fibers running from the cortex to the cochlea (Békésy, 1967), parallel to the afferent auditory fibers.

The extent of our knowledge of the role of these efferent tracts is still extremely limited. It would appear, however, that they exert a controlling influence over the status of the afferent system. This is quite a startling concept, for it requires that we reject the idea of the auditory pathway as a passive system for conveying information upward. Instead, we must begin to consider it as an active system, subject to tuning effected by higher centers. Experiments (Desmedt, 1960, 1962) have indicated that in animals the inhibitory effect of the efferent system, when stimulated electrically at the midbrain level, suppresses the response of the auditory nerve to acoustic clicks (short impulse sounds). The equivalent threshold change (reduced sensitivity) was of the order of 20 to 25 dB. Other workers (Whitfield and Comis, 1967) have shown that *increased* activity in neural centers of the auditory pathway can be evoked by stimulation of the efferent tract at the level of the brain stem. It was demonstrated that stimulation of the olivary complex effectively lowered (improved) the sensitivity of the cochlear nucleus, thus enhancing the sensitivity of the auditory system to sound stimulation. Other researchers have reported the effects of efferent tuning at the level of the eighth nerve (Katsuki, 1961; Nomoto et al., 1964). It would appear, therefore, that the efferent fibers are able to control the flow of afferent impulses, either enhancing or decreasing responses of neuron groups. This process of attenuating certain impulses, thus effectively enhancing others, is known as *lateral inhibition:* By means of this feedback loop the auditory system (efferent and afferent fibers) is rendered self-regulating. Switching, or *gating*, of impulses permits control of the patterns of stimuli allowed to pass upward to higher-order neurons. It may even make possible the gating of input-output flow at lower levels, which would seem to be substantiated by experiments in animals where certain auditory response behaviors are maintained even when higher centers are surgically destroyed (Neff, 1961; Masterton and Diamond, 1964; Cornwall, 1967).

When we discuss theories of speech perception in Chapter 5, we will refer again to the possible role of the efferent system in tuning the afferent pathways so that particular stimuli can be identified in the complex pattern arriving at the cochlea. The presence of the efferent system certainly adds support to the concept of the afferent sensory system as a dynamic system which, like the other sensory systems, is actively involved in the processes of perception.

Brain-stem Nuclei

Although the roles of the specific nuclei in the auditory pathway have not been determined, some general information has been obtained. The cochlear nuclei are known to be important for primary levels of auditory discrimination involving frequency and intensity analysis. The nerve fibers entering the cochlear nuclei have been shown to preserve the anatomical distribution of the fibers in the auditory nerve, as well as the tonal arrangement of frequencies across each nucleus (tonotopic organization). At the level of the superior olive the first interaction of fibers between the two ears occurs, providing for comparison of binaural data. Whitfield suggests that this nucleus is therefore concerned with processing information pertaining to directionality.

The nucleus of the lateral lemniscus is generally considered as an extension of the olivary complex. Tonotopic organization appears to be less defined than in the cochlear nucleus, but there begins to occur a phenomenon which becomes progressively more evident at the higher levels of the auditory pathway: namely, selectivity of response to stimuli, part of the gating already referred to. Intensity differences of as much as 50 dB have been shown between the pure tone thresholds for different cell units (Galambos, Schwartzkopff, and Rupert, 1959). Some cells appear to respond only to stimuli from the contralateral ear, while others respond to signals from either ear (Hall, 1964). The cells of the olivary complex and lateral lemniscus have been shown to be highly sensitive to the time characteristics of the signal.

The inferior colliculus also has tonotopic organization and is thought to play an important role in tone discrimination. The medial geniculate body, on the other hand, does not appear to be tonotopically organized nor does it seem necessary for tonal discrimination.

The impression is that up to the level of the inferior colliculus the neural structures are exclusively concerned with the processing of auditory data. The interconnecting neurons below the colliculus run only between structures which are equally specific in their sensitivity; that is, they are responsive only to acoustic stimuli. However, at the level of the medial geniculate body this exclusive sensitivity begins to break down, with parts of the nucleus no longer specifically concerned with auditory information. Once the cortex is reached, the interconnections between areas of different types of sensitivity become extensive.

In outlining and explaining the model of brain functioning developed by Bronson (1965), John Nash discusses these interconnecting neurons found within each of the levels of the cortex. He states:

> Within each of these levels there is a network of short axon neurons with multiple interconnections, and also long axon neurons that interconnect regions within the level. Other long axon neurons connect the levels both upward and downward. There are in addition the peripheral afferents and efferents that connect the sense organs with the CNS at the various levels allowing increasingly refined and successively more differentiating, sensory and motor discriminations to be made by the networks. The long axon neurons for vertical integrations between the levels permit the more primitive systems to exercise an upward control on the programming of patterns of cerebral activation. Those projecting downward from the higher levels impose tonic inhibition and more selective phasic excitations and inhibitions on the lower level systems, thus differentiating overall functions. There is an interesting diversity of neural interconnections within the higher levels in more advanced species and this diversity allows for an increasing development of specialized functions in the short axon networks. (1970, p. 117)

The Auditory Cortex

The auditory cortex is located in the temporal lobes of the brain. There appears to be a primary area and a secondary area. The exact boundaries of these two areas have not been defined because as one moves away from the small area, which is almost unequivocally concerned with the processing of auditory information, the exclusiveness becomes less and less apparent. There is no clear boundary for those cells dealing with auditory data, but a progressive involvement in responsiveness to different types of data processing. Whitfield has summarized this characteristic of the auditory cortex:

> The "auditory cortex" is hard to define, either anatomically or functionally. While certain areas are so predominantly connected with the rest of the auditory system as to be unequivocally auditory areas, surrounding regions become less and less certainly so, as we move farther and farther from the primary projection area. (1965, p. 123)

It is of interest to note that even the secondary auditory areas are not entirely dependent upon activation by the primary areas, but receive some impulses from the level of the cochlear nucleus (Downman, Woolsey, and Lende, 1960).

In addition to the primary and secondary auditory areas, other regions of the brain have been demonstrated to be responsive to stimulation of the auditory nerve. It appears, therefore, that a number of areas, not exclusively auditory, receive and process auditory information. Conversely, it has been reported (Evans and Whitfield, 1964) that primary auditory areas contain nerve cells which are responsive to visual stimulation (2%), while a lower percentage (0.5%) were responsive to both auditory and visual stimulation. Lenneberg (1967, p. 55) has stated that with the exception of part of the occipital lobe, which is known to be the primary visual area, *"There is no other area in the human cortex which is both histologically distinct and unequivocally and uniquely related to one and only one motor or sensory function."* (italics mine)

The role of the auditory cortex appears to be the recognition of complex patterns in auditory stimuli. It seems improbable that it is in any way concerned with recognizing the absolute measures of frequency and intensity (Goldstein, 1961; Hodgson, 1967) but rather with determining the relationship of these two parameters over time. The capacity to analyze the temporal components is referred to as the *temporal resolving power* of the auditory system. It is likely that the recognition of complex sounds is dependent upon cortical processing, while lower-order neurons are involved in analyzing the frequency/intensity components of the complex wave. Thus, the cortex identifies the sequential patterning essential to speech perception, apparently stores electrical-impulse patterns on a "hold" basis (Whitfield, 1967), and performs other complex pattern-synthesis tasks.

Summary

In summary we can say the following:

Knowledge of the structure and function of the auditory system is important because it helps us realistically to develop and amend our theories of perception. Theories, to be effective, must grow out of a judicious blending of known facts, logical deductive and inductive reasoning, and a sprinkling of speculation.

We have examined the anatomical structures of the auditory system in order to obtain some insight into how detection and replication of the acoustic pattern occurs. We learned that from the eardrum to the cilia of the hair cells the process is either mechanical or hydraulic. However, once the hair cell has been stimulated to fire an impulse in the acoustic nerve fiber, a component of the pattern has been internalized. The major difference between the system up to the hair cells and the system beyond them to the cortex is that the first section is purely conductive, while beyond the hair cells it is both conductive and analytical.

Our examination of the neural pathway has revealed that the auditory system becomes increasingly complex as it moves to higher levels in the brain. The number of cells in the cochlear nucleus of a monkey has been estimated at 88,000 (Chow, 1951); the number of these cells in the human is by contrast almost beyond comprehension! In addition to the increasing number of cells, the number of connections between cells and between neural structures increases enormously as the pathway moves upward toward the cortex. We have considered evidence that efferent fibers running alongside the afferent tracts are able to influence the behavior of the sensory neurons, opening pathways which would otherwise be blocked, increasing and decreasing the sensitivity of groups of neurons. This interaction of efferent and afferent fibers provides a feedback loop by which the system can control its status, tuning itself according to the needs of the organism for certain types of auditory processing.

Finally we saw that the specificity of the system so apparent at the lower levels begins to break up at the level of the medial geniculate body.

At this point tonotopical arrangement is no longer present; only part of the nucleus is specific to auditory stimulation, and many connections are made to nonauditory nuclei. This process is even more pronounced at the cortical level, where the boundaries of the primary areas become hard to define as cell specificity begins to blur and as areas are concerned with more than one type of stimulus.

The auditory pathway is, then, far more intricate and involved than one might at first suspect from introductory texts on hearing. However the crux of our discussion lies in the mounting evidence which seems to support our concept of the sensory systems as dynamic, self-regulating systems actively searching for *informative patterns* within the complex of information potentially available in the media surrounding them.

REFERENCES

BATTEAU, D. W., 1967. The role of the pinna in human localization. Proceedings of The Royal Society, Series B., **168**, No. 11011, 158–180.

BAUER, R. W., J. L. MATUZSA, R. F. BLACKMER, AND S. GLUCKSBERG, 1966. Noise localization after unilateral attenuation, *J. Acoust. Soc. Amer.*, **40**, 441–44.

BÉKÉSY, VON G., 1967. *Sensory Inhibition.* Princeton, N.J.: Princeton University Press.

BÉKÉSY, VON G., 1960. *Experiments in Hearing.* New York: McGraw-Hill.

BÉKÉSY, VON G., 1947. The variation of phase along the basilar membrane with sinusoidal vibrations. *J. Acoust. Soc. Amer.*, **19**, 452–60.

BÉKÉSY, VON G. AND W. A. ROSENBLITH, 1958. "The Mechanical Properties of the Ear," in *Handbook of Experimental Psychology*, ed. S. S. Stephens. New York: Wiley.

BRONSON, G., 1965. The hierarchical organization of the central nervous system: implication for learning processes and critical periods in early development. *Behav. Sci.*, **10**, 7–25.

CHOW, K. L., 1951. Numerical estimates of the auditory central nervous system of the rhesus monkey. *J. Comp. Neurol.*, **95**, 159–75.

CORNWALL, P., 1967. Loss of auditory pattern discrimination following insular-temporal lobe lesions in cats. *J. Comp. Physiol. Psychol.*, **63**, 165–68.

DAVIS, H., 1962. Advances in the neurophysiology and neuroanatomy of the cochlea. *J. Acoust. Soc. Amer.*, **34**, 1377–85.

DAVIS, H., 1960. "Mechanism of Excitation of Auditory Nerve Impulses," in *Neural Mechanisms of the Auditory and Vestibular Systems*, eds. G. L. Rasmussen and W. Windle. Springfield, Ill.: Charles C Thomas.

DESMEDT, J. E., 1962. Auditory evoked potentials from cochlea to cortex as influenced by activation of the efferent olivocochlear bundle. *J. Acoust. Soc. Amer.*, **34**, 1478–96.

DESMEDT, J. E., 1960. "Neurophysiological Mechanisms Controlling Acoustic Input," Chapter 11 in *Neural Mechanisms of the Auditory and Vestibular Systems*, eds. G. L. Rasmussen and W. F. Windle. Springfield, Ill.: Charles C Thomas.

DOWNMAN, C. B., C. N. WOOLSEY, AND R. A. LEADE, 1960. Auditory areas I, II, and Ep: Cochlear representation, afferent paths and interconnections. *Johns Hopkins Bulletin*, **106**, 127–42.

EVANS, E. R. AND I. C. WHITFIELD, 1964. Classification of unit responses in the auditory cortex of the unanaesthetised and unrestrained cat. *J. Physiol.*, **171**, 476–93.

FREEDMAN, S. J. AND H. G. FISHER, 1968. The role of the pinna in auditory localization, Chapter 8 in *The Neuropsychology of Spacially Oriented Behavior*, ed. S. J. Freedman. Homewood, Ill.: Dorsey Press.

GALAMBOS, R., J. SCHWARTZKOPFF, AND A. RUPERT, 1959. Microelectrode study of superior olivary nuclei. *Amer. J. Physiol.*, **197**, 527–36.

GOLDSTEIN, R., 1961. Hearing and speech in follow-up of left hemispherectomy. *J. Speech and Hearing Dis.*, **26**, 126–29

HALL, J. L., II, 1964. Binaural Interaction in the Accessory Superior Olivary Nucleus of the Ear. Technical Rep't. 416. Research Laboratory of Electronics. Cambridge, Mass.: The M.I.T. Press.

HARRIS, J. D., 1974a. *Anatomy and Physiology of the Peripheral Hearing Mechanism*. The Bobbs-Merrill Studies in Communicative Disorders. Indianapolis and New York: The Bobbs-Merrill Co.

HARRIS, J. D., 1974b. *The Electrophysiology and Layout of the Auditory System*. The Bobbs-Merrill Studies in Communicative Disorders.

HELMHOLTZ, H. L. G., 1863. *On the Sensations of Tone*, 3rd Edition, trans. A. J. Ellis. London: Longmans Green.

HODGSON, W. R., 1967. Audiological report of a patient with left hemispherectomy. *J. Speech and Hearing Dis.*, **32**, 39–45.

KATSUKI, V., 1961. "Neural Mechanism of Auditory Sensation in Cats," in *Sensory Communications*, ed. W. A. Rosenblith. Cambridge: M.I.T. Press, 561–84.

LENNEBERG, E. H., 1967. *Biological Foundations of Language*. New York: Wiley.

LORENTE, DE Nó, R., 1933. Anatomy of the eighth nerve III, General plan of structure of the primary cochlear nuclei. *Laryngoscope*, St. Louis, **43**, 327–50.

MASTERTON, R. B. AND I. T. DIAMOND, 1964. Effects of auditory cortex ablation on discrimination of small binaural time differences. *J. Neurophysiol.*, **27**, 15–36.

NASH, J., 1970. *Developmental Psychology: A psychological approach*. Englewood Cliffs, N.J.: Prentice-Hall.

NEFF, W. D., 1961. "Neural Mechanisms of Auditory Discrimination, in *Sensory Communciations*, ed. W. S. Rosenblith. Cambridge: The M.I.T. Press, pp. 259–78.

NOMOTO, M., N. SUGA, AND Y. KATSUKI, 1964. Discharge pattern and inhibition of primary auditory nerve fibers in the monkey. *J. Neurophysiol.*, **17**, 768–87.

RASMUSSEN, G. L., 1960. "Efferent Fibers of the Cochlear Nerve and Cochlear Nucleus," Chapter 8 in *Neural Mechanisms of the Auditory and Vestibular Systems*, eds. G. L. Rasmussen and W. F. Windle. Springfield, Ill.: Thomas.

RASMUSSEN, G. L., 1958. Anatomical discussion of neural mechanisms in audition, by Galambos, R., *Laryngoscope*. St. Louis, **68**: 404–406.

RASMUSSEN, G. L., 1955. Descending or 'feedback' connections of the auditory system of the cat. *Amer. J. Physiol.*, **183**, 653.

RASMUSSEN, G. L., 1953. Further observations of efferent cochlear bundle. *J. Comp. Neurol.*, **99**, 61–74.

RASMUSSEN, G. L., 1946. The Olivary Peduncle and Other Fiber Projections of the Superior Olivary Complex. *J. Comp. Neurol.*, **84**, 141–219.

RASMUSSEN, G. L., 1942. An efferent cochlear bundle. *Anat. Rec.*, **82**, pp. 441.

RHODE, W. S., 1971. Observations of the vibration of the basilar membrane in squirrel monkeys using the mossbauer technique. *J. Acoust. Soc. Amer.*, **49**, 1218–31.

RUTHERFORD, W., 1886. A new theory of hearing. *J. Anat. Physiol.*, **21**, 166–68.

SPOENDLIN, H., 1966. "Structural Basis of Peripheral Frequency Analysis," in *Frequency Analysis and Periodicity Detection in Hearing*, eds. R. Plomb and G. F. Smoorenburg. Leiden: Sijthoff, pp. 2–36.

WEVER, E. G., 1949. *Theory of Hearing*. New York: Wiley.

WEVER, E. G. AND M. LAWRENCE, 1954. *Physiological Acoustics*. Princeton: Princeton University Press.

WHITFIELD, I. C., 1967. *The Auditory Pathway*. London: Edward Arnold, Ltd. (Publisher), pp. 27–28.

WHITFIELD, I. C., 1965. "Edges" in auditory information processing. XXIII International Physiol. Cong. (Tokyo): 245–47.

WHITFIELD, I. C. AND S. D. COMIS, 1967. "A Reciprocal Gating Mechanism in the Auditory Pathway," in *Cybernetic Problems in Bionics*, eds. H. L. Oestreicher and D. R. Moore. New York: Gordon and Breach.

ZEMLIN, W. R., 1968. *Speech and Hearing Science*. Englewood Cliffs, N.J.: Prentice-Hall.

4

Fundamentals of
Pattern Processing

The situation of a person sitting down to write a chapter concerned with the processes involved in perception is analogous to David setting out to slay the giant; it is a monumental task not to be undertaken lightly. Yet one cannot but help reflect upon the fact that David was successful. He achieved his goal because he was able to concentrate all his efforts upon a critical area of the giant's body. In this chapter we shall attempt to follow his example by focusing our efforts upon that notion which appears to be critical to an understanding of speech perception. Our focus will center upon the *recognition of patterns* by the perceptual system. The structure of the acoustic speech signal and the organization of the auditory system lend credence to the idea that the ability to classify patterns is fundamental to auditory perceptual function. Indeed, evidence abounds to suggest that it is fundamental to all perceptual systems.

Throughout our discussion of the acoustic signal it was stressed that the parameters of intensity and frequency interact over time to generate various patterns of complex speech sounds. Each pattern can be identified on repeated presentation by means of *distinctive features*; i.e., those aspects of the complex stimulus which appear to be directly correlated to the individual's consistency of perception of that stimulus. Whether indeed there is something like a perceptual unit value *within* the stimulus (Ladefoged,

1959; Miller and Isard, 1963) which triggers the percept in the person, whether the percept arises from information generated by the pattern of organization of the sensory-perceptual system, or whether it is in some way related to the articulatory production of the sound will be considered in the next chapter. Whichever is the case, the auditory system, from the cochlea to the cortex, is concerned with two tasks: (1) *analyzing* the variations in the separate dimensions of intensity and frequency over a period of time, and (2) *resynthesizing* those variations to develop an internal representation of the pattern. The task confronting us in this chapter is to examine some of the fundamental processes underlying pattern recognition in general, so that we can apply them to speech perception in the next chapter.

We know that our sensory end organs comprise an interface between us and the physical world. The only thing that can flow across that boundary between the physical world and our sensori-perceptual system is information. The effect of information is to constrain the organism in the manner in which it can internally restructure the external pattern. Each successive datum reduces uncertainty and therefore limits the number of possible patterns, just as each correctly placed piece of a jigsaw puzzle serves to guide one's selection of the next.

How information is extracted from a physical stimulus is a key question. As in solving a jigsaw puzzle, it is apparent that this process depends upon the identification and categorization of certain characteristics of each piece. The final selection of a particular piece is possible because it differs sufficiently from all other pieces and therefore can be recognized as a "fit." The recognition of these distinctive features permits us to select what is probably the most appropriate piece and to reject all others. Our prediction must then be verified by actual proof as additional pieces are added. A similar categorization process is thought to occur in speech processing.

Categorization of Stimulus Features

Most theories of pattern perception assume that categorization occurs as a function of a *binary system*; that is to say, each decision results in the rejection of one of two equally probable categories on the basis of a yes/no answer. Information is measured in terms of how many such questions must be asked before a positive identification of a stimulus category can be made. For example, assume you must identify one of four persons who vary by sex and/or color of their eyes. You know that one of the males and one of the females has brown eyes, while the other two are blue eyed.

In order to identify positively one of these four persons by the criteria

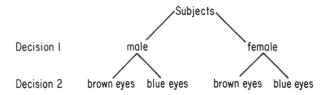

Figure 4.1   A two-decision identification process on a binary system.

of sex and color of eyes, you will need to make two decisions on a binary choice basis:

1. Is the subject male or female?
2. Is the subject blue or brown eyed?

Now substitute for the people four acoustic stimuli with the characteristics of frequency and intensity. The categories of acoustic characteristics which produce the stimulus patterns are therefore:

High frequency
Low frequency
High intensity
Low intensity

The two decisions which must be made to identify the stimulus appear in Figure 4.2.

In situations where the occurrence of either one of two characteristics is equally probable, and when one decision permits stimulus identification, the pattern is said to carry one *bit* of information.

In Figure 4.3 we observe that in order to identify a particular component in (a) only one decision needs to be made; (b) necessitates the separation of the initial two categories, each with two equally probable char-

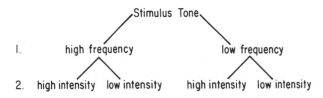

Figure 4.2   A two-decision, binary system identification of a complex tone.

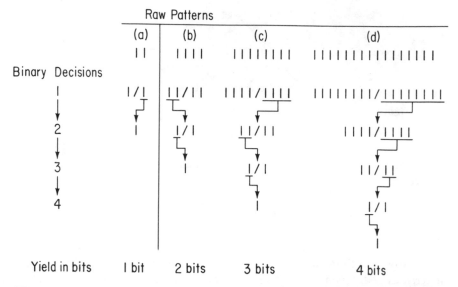

Figure 4.3 Information yield as a function of the number of characteristics.

acteristics and then requires a second decision to establish which of the remaining two possible characteristics is in fact present. In (c) the process requires three such decisions. If sixteen characteristics are present as in (d), four decisions will yield identification.

The concept of information, and the manner in which it is processed is relevant to our discussion. Many research findings pertaining to the identification of stimulus patterns have been derived from experiments based upon information theory. The underlying philosophy is that pattern recognition is significantly affected by the observer's concept of the total set of patterns which he might encounter. It has been shown that the amount of information which can be extracted from a complex stimulus is proportional to the degree of uncertainty it represents. That is to say, if a person is aware of all the possible combinations and, most importantly, of the probabilities of the occurrence of each combination, he will quickly recognize patterns. This situation describes the rapid speech perception of a person well acquainted with a particular set of language rules. We are especially interested in the identification of probabilities by the observer and the subsequent tuning of the system to increase its sensitivity to those stimuli predicted as relevant to the current perceptual needs of the person. The number of bits of information a person needs to identify a specific stimulus pattern is not absolute for that stimulus pattern. Different individuals may require significantly different amounts of information to identify the same stimulus. Even the same person may require different amounts on different occasions. We can measure this by determining the number of bits of information which pass through the perceptual system to evoke a response.

We are concerned, therefore, with a process involving the restructuring by the nervous system of information patterns which have activated our sensori-perceptual system. As Neisser has stated:

> The central assertion is that seeing, hearing, and remembering are all acts of construction, which may make more or less use of stimulus information depending upon circumstances. (1967, p. 10)

The particular modes, or *strategies*, employed to analyze the information conveyed by the physical stimulus have been subject to much speculation and study. Since these concepts are important to the theoretical models of speech perception we shall consider later, we will devote a few pages to discuss the basic ideas now.

Processing Strategies

A dynamic sensori-perceptual system, which can adapt to the demands placed upon it by tuning or focusing in a selective manner, must be able to search effectively the vast amount of stimulus information which is activating its peripheral sensory organs. Once the desired pattern is identified, the question is whether the selected stimulus complex is then analyzed in a step-by-step process (*serial processing*), or whether there is a simultaneous examination of the features of the stimulus (*parallel processing*).

Serial processing   Figure 4.4 illustrates serial processing. A pure tone of high or low frequency, high or low intensity, and of long or short duration, enters the analyzer. The tone is one of eight, each with a different combination of features. The analyzer seeks to identify it. The first feature-detector determines the highness or lowness of frequency, the second determines whether the stimulus has high or low intensity, while the third must define the stimulus as of long or short duration. As each analyzer produces its specific data, the results are fed to storage. Only when the final analyzer (n) releases its findings does the response-determiner request the release of stored information.

Since the speech stimulus is a transitory one, the time factor becomes critical in serial processing, for each feature-analyzer added to the system prolongs the process. While it is possible to envisage a smoothly functioning serial system capable of rapidly analyzing a complex stimulus with rela-

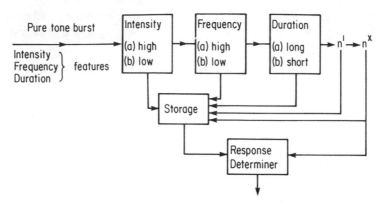

Figure 4.4   Serial processing of data.

tively few distinctive features, it is hard to imagine this process accommo-
dating a stimulus with multiple features requiring rapid analysis. Furthermore, a serial system requires a storage component. Since the stimulus is
restructured as a series of steps, the results of the early stages must be
placed on "hold" until the constraints of the later stages are determined.
The resultant time delay becomes increasingly greater as additional feature-
detection units are added to cope with increasingly complex patterns containing multiple components. When the storage requirements grow large,
the potential for a fading of stored data becomes a critical factor. The pattern reconstruction may then be distorted because of missing or deteriorated data incorrectly analyzed at the early stages.

　　This type of process is also limited by the substantial effect an incorrect analysis of a given component may have upon pattern perception.
When subsequent processing occurs under the influence of data already perceived, the potential for error is great; of course, the influence of error on
the final percept is greatest when it occurs during the early stages. This
may seem, therefore, to be an inappropriate way for the perceptual systems to process sensory data, yet the evidence (Neisser, 1963) suggests that
this manner of scanning is in fact utilized in responding to new patterns of
stimulation. This type of processing probably underlies the slow response
behavior of young children attempting new tasks. It is a cumbersome way
to do things but appears to be a necessary component of new learning.

　　Parallel processing   As an individual becomes familiar with a task, he
apparently organizes his system to operate in the more efficient parallel
mode. Parallel processing (Figure 4.5) involves the simultaneous examination of all distinctive features. It overcomes the problem posed by increasing reaction time in the serial system. It also reduces the impact of each
feature-analyzer on the final percept, thus minimizing the error, and it vir-

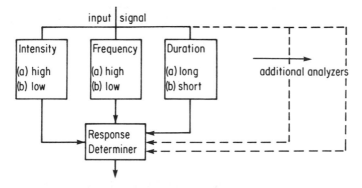

Figure 4.5   Parallel processing of data.

tually eliminates storage effects. Figure 4.5 illustrates how the complex signal is simultaneously presented to the feature-analyzers which then synchronously feed their results to the response-determiner where the pattern is synthesized and identified. Analyzers arranged in parallel function independently. This means that highly complex patterns can be processed as rapidly as simple ones. Theories of pattern identification have utilized either or both of these approaches, but emphasis has increasingly been placed upon the use of parallel models.

Theories of Pattern Identification

To reiterate, the question confronting us in attempting to understand speech perception is how information (i.e., identifying features of the complex pattern imposed upon the sound wave by the human articulatory/resonance system) can be restructured internally to yield perceptual value. In this chapter however, we shall merely examine several models applicable to all forms of perception.

The development of these general models, and the ability to determine their internal validity, has been greatly aided by the advent of the computer. There are numerous reasons why scientists are interested in developing machines which can perform human functions. Experimental machines can already read aloud the printed word, they will soon no doubt be able to translate languages, type to dictation, and perhaps even carry on a conversation. One can hardly label this forecast science fiction when one considers the increasingly effective use of technology in monitoring tasks requiring rapid decision making.

In order to construct machines capable of processing data as efficiently as the human perceptual system we must understand how the human nervous system is capable of identifying patterns. How can physical stimuli which differ so markedly in their pattern organization upon each occurrence be perceived as constant images? The computer has come to be rather like a consultant in this search. It serves two major purposes. First it provides a means of testing theories to see if they can *really* be made to work. This enormously valuable function allows the theorist to demonstrate that a given structure would actually function as claimed if it could be built. The theoretical model can thus be proved valid not only logically but also practically when the computer is able to perform those tasks the model attempts to explain.

The second reason for turning to the computer is that it is capable of indicating at which point in a programmed model a problem occurs. This makes it easier for the theorist to make modifications in the model. Of course, even a machine built to "understand" and react appropriately to speech will not necessarily replicate the manner in which humans understand and react appropriately to speech; there are many ways to skin a cat. Nevertheless, when a theory is developed within the constraints of known facts and is then tested by the computer, the designer can at least have faith that he has devised a model which accurately describes *one* way in which the human system might work.

Rather than going into the history of discarded models of pattern recognition, we shall confine our discussion to those models that permit us to bridge the gap between the material studied in Chapters 2 and 3 and the specific theories of speech perception we will consider in the next chapter.

Theories of pattern recognition can be classified into four categories:

1. template theories
2. filtering theories
3. feature-detection theories
4. analysis-by-synthesis

Template Theories   Theories in this category approach the task as one of matching, or correlating, input pattern information to an internal standard pattern—a type of pattern bingo. Such a system operates as does a bank clerk in matching the signature on a withdrawal form with the original signature kept on file. The teller will accept the withdrawal only if the pattern of handwriting remains essentially within the boundaries of his or her visual perceptual sets for "a match."

We shall not consider the template-matching models because they do not seem to be easily applied to the demands made upon a model by the process of speech perception. This theory has difficulty in allowing for the wide variations which occur in the production of speech-sound patterns by a given speaker upon repeated occasions, or between speakers with different speech patterns. Although some attempts have been made to explain speech perception by this theory (Arbib, 1964), there is another factor mediating against it. Our perceptual system appears to be able to attribute phonemic value to nonspeech sounds, providing certain acoustic constraint criteria are also present (Liberman, Delattre, and Cooper, 1952). Clearly, templates cannot exist for nonspeech sound/noise bursts.

Filtering Theories   In these theories information is envisaged as passing through banks of filters which permit the sorting of the data into identifiable categories. Fant (1967) has used filtering to explain the process of speech perception; other writers have used it to explain the process of selective attention, introducing the idea of variable rather than fixed filter characteristics. The filter theory has recently been radically modified to allow for the concept of active neural units which select rather than filter. This alternative explanation is known as the feature detection theory.

Feature Detection   This approach to pattern recognition envisages a system comprised of discrete detectors, each sensitized to a certain potential pattern component(s) or feature(s). The pattern is reconstructed in terms of a feature analysis of the input signal. The models based upon this theory posit individually sensitized feature-detection units—in the human system, neurons or patterns of neurons which are activated when a component or combination of components within the complex pattern coincide with their response criteria. In some of the models, the flow of information upward through the system occurs in the form of a continuous electrical analogy (*analogue*) of the patterns represented by the feature detectors. Other models use a *digital system* in which the information is represented by a series of discrete pulses representing yes/no answers to discrete questions: a binary code. When a feature-detection unit is triggered in analogue models, it passes upward a blueprint of the combination of criteria for which it stands. In these models the restructuring of the total pattern, i.e., its perception, occurs only at the highest level. At this level all the individual decisions can be brought together to reproduce a copy of the recorded input pattern. The pattern will then be perceived in terms of the aggregate of the individual unit anlogues. It is important to note that the input data will be perceived, not in terms of the external stimulus, but in terms of how the system has processed them. The processing is subject to both internal and external influences. Aside from these, the peripheral data undergo numerous transformations. To quote Berry:

> By the time the temporal spacial patterns reach cortical circuits, the code may bear little relation to the original stimulus pattern. Indeed,

perception may be so altered by the reticular system, or by the earlier processes of coding that it represents a distortion of the actual nature of the stimulating world. (1969, p. 101)

Or in the same vein:

What the subject sees then is not constructed from the units of the stimulus, as a wall is made from bricks; rather what the subject sees represents the results of encoding routines or cognitive operations that work on whatever materials are available. (Kolers, 1967, p. 224)

The most fascinating illustration of the feature-detection type of model is the one proposed by Selfridge (1959). The model (Figure 4.6) is intriguingly presented in anthropological terms. He gave his model the unusual title of "Pandemonium." It envisages a parallel system at each level of processing. The model involves several transformational stages culminating in a decision-making component. At the first level (a) a set of *data or image "demons"* store the input pattern and pass a copy up to the second order (b) *computational demons* who are concerned with analysis of the data. Applied to the auditory system, each demon at this level would compute the data to provide information concerning the spacial and/or tem-

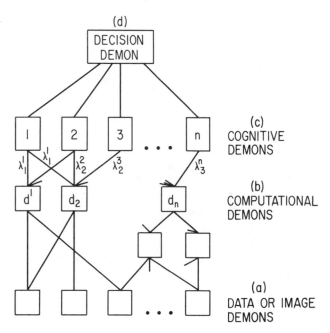

Figure 4.6   Pandemonium Model. (after Selfridge, 1959).

poral components of the acoustic data. It is at this level that information would be generated concerning the frequencies present, the concentration of energy within certain frequency ranges, the changes of frequency arrangement both in terms of direction (rising or falling) and rate, intensity variations, durational variables, etc.

The third order (c) *cognitive demons* are an egotistical group, who constantly search the computed data for any resemblance of their own characteristic features. These are the feature-detectors of the model. When a computational demon does detect a reflection of himself in the input data he *"shrieks"* the fact to the superior (d) *Decision Demon*. The closer the match between the input pattern and his own particular features, the more delighted he is and the louder he shrieks. The Decision Demon is faced with all, or at least most, of the cognitive demons, shouting at varying levels of loudness. Utter "pandemonium"! Rather like the classroom teacher who, having asked a question, is faced by a horde of little demons all jumping up and down for attention. She may, as the Decision Demon does, assume that the most excited child probably has the right answer. The Decision Demon accepts the loudest shriek as the correct response.

The pandemonium model is offered as a dynamic system. First, it allows for tuning, or as we have referred to it earlier, "gating," by permitting a weighting of the value of the shrieks from each cognitive demon. This provides for the tuning of the system to be most sensitive to those features in the input pattern which are predicted as being relevant to the needs of the organism. The weighting (tuning) can therefore be modified by rearranging the relationship pattern between the cognitive and computational demons on the basis of the effectiveness of a particular weighting on a previous experience. This ties in very nicely with the neurological concepts we discussed in the last chapter (pp. 71–72) and concurs with Berry's (1969) statement which she supports by reference to Adrian (1947):

> The central processing of sensory information, however, cannot be measured by peripheral input. Information not useful to the organism at the moment will be inhibited by the alerting mechanism, so that only sensory patterns having significance for the subject will be admitted to cortical analysis. Cortical discrimination or perception seems tied to facilitatory and inhibitory activity of brainstem systems. (1969, pp. 40–41.)

In another model Uttley (1959, 1966) has proposed a modification of the feature-detection approach of Selfridge. In the Selfridge model, the decision is made on the basis of the activity of simultaneously functioning feature-detection units. Uttley, whose model we shall discuss in more detail in the next chapter, conceives rather of a hierarchy of increasingly sophisti-

cated feature-detection units, each having increasingly complex requirements for firing. While at the periphery a single component of a pattern
may trigger a neuron, at the next higher level a combination of two criteria
may need to be met before a unit will fire. Neurons which are activated
only by a multidimensional stimulus are referred to as *indicator units*, since
the firing of such a unit indicates which lower-order units must have triggered its response. The indicator unit comes to represent a pattern of features which may be quite simple at the lower levels but can become
increasingly complex as higher level indicators require the integration of
two or more simple patterns to trigger them. This represents a series of
transformations of simple raw data entering the system at the periphery
into a coded representation of the interaction of multiple, individual features into a complex pattern. The concept of indicator units, which represent coded pattern values, fits the evidence that in the auditory system the
peripheral neurons are highly specialized, while at higher cortical levels the
specificity progressively decreases as the cells begin to be responsive to
multiple inputs.

Analysis-by-Synthesis   This model, proposed by Kenneth Stevens
and Morris Halle (1967), is designed to explain in particular the manner
in which speech sounds are identified. For this reason we will defer a careful discussion of this theory until the next chapter. The authors point out
that the ideas basic to the concept of analysis-by-synthesis have their roots
in the rationalist theories of perception. They postulate that:

> . . . the perception of speech involves the internal synthesis of pat-
> terns according to certain rules and a matching of these internally
> generated patterns against the pattern under analysis. (1967, p. 88)

The analysis-by-synthesis model involves, therefore, the identification of
patterns predicted on the basis of the constraints operating at a given
time. Under the constraining influence of the rules stored in the system,
an internal pattern is generated. The pattern is a computed probability
—an expectancy. This constitutes the synthesis stage. Analysis is then
performed by comparing the *predicted* signal with the received input of
the *actual* signal. Solley and Murphy (1960, p. 156) have suggested a similar stage in their perceptual model. They conceive of a "trial and check"
stage following what they term an attention stage.

In general, theorists have felt that the perceptual process is most
likely to operate on a system which is sensitive to the temporal and spacial
variations of stimuli.

It is a basic principle in human sensory functions that we are especially sensitive to varitions of stimuli in place and time. We should pay more attention to this principle if we want to approach the problem of basic dimensions of perceived sound patterns and thus approach a dynamical theory of speech perception. (Fant, 1967, p. 119)

It follows, therefore, that feature-detection theories involving the concept of a hierarchical arrangement of neural feature-analyzers are likely to increasingly dominate approaches to perception, and in particular to speech perception. However, it is not only necessary to discover which distinctive features in the pattern serve to permit its internal restructuring; it is also important to discover how the organism selects a desired pattern from the multitude of stimuli potentially available to it. While the organism has no way, at least in hearing, of preventing the auditory stimuli from reaching the sensory nerve endings, it must be selective in what it processes if it is to avoid being totally overwhelmed. Therefore, before we can progress further with our consideration of speech perception, we will need to examine the basic concepts of selective attention.

Selective Attention

In our discussion of the auditory system, we paid particular attention to the role of the efferent pathway in attuning the system to those patterns which are anticipated as relevant to the needs of the organism. This selectivity of the feature-detection units has been reviewed by Solley and Murphy in their work on selective attention (1960). They point out that "The organism must 'select'; or to speak more precisely, some sort of competition among the various perceptual tendencies must be resolved by a priority system within the living structure itself." (p. 178) The authors consider the process of attending to be biological, consisting of a series of transitions from one response to another. They see selective attention as involving the adjustment of sense organs (p. 180). This view is highly compatible with the concept of gating, or sensori-neural tuning, which we have already considered. It also concurs with Gibson's view that attention is "a way of orienting the perceptual apparatus of the body" (Gibson, 1966). There is a growing amount of evidence from the field of neurology, that during the act of attention efferent fibers exert considerable inhibitory influence on the neuronal activity in the neural cells surrounding the focus of the excitation:

There is strong evidence that this descending inhibitory influence may play a role in "editing" the flow of information by acting to suppress some of the input from the periphery of the receptive field and thereby producing an effective inhibitory surround to the main focus. (Brazier, 1964, p. 1427)

Tuning thus becomes a question of variable sensitivity of certain neural groupings, rather than a situation in which cell groups either fire or are inactive. This fits with the theoretical behavior of Selfridge's cognitive demons, who all shriek to varying degrees in response to a complex stimulus, but are weighted depending upon the anticipatory set of the system. It seems logical to assume that when we speak of attention, some form of relative scaling of the activity of neural groupings is involved. We are not dealing with an act of attention which involves a complete blocking out of all available stimuli except those in which we are interested, but rather with an hierarchical ordering of available stimulus pattern information, an ordering that can be modified rapidly as needs change.

Solley and Murphy refer to, and quote from, many authors who consider attention to constitute a single process involving the relative distribution of a finite source of neural energy. They refer to the work of Stern who writes:

Since the person's supply of energy is limited at all times, concentration upon one area is obtained at the cost of withdrawing energy from others. (1938, p. 475–476)

Neisser (1967) rejects this view. He maintains that "attention is not a mysterious concentration of psychic energy: it is simply an allotment of analyzing mechanisms to a limited region of the field" (p. 88). It is his opinion that selective attention is an aspect of information processing. This has been our approach to perception. It also concurs with Solley and Murphy's view that attention is a biological process. Neisser suggests that we attend selectively by processing the information from the object of our attention in a different, more sophisticated manner than the information from other sources of potential stimuli. Selective attention thus becomes *a function of the manner of processing of the information.* He argues that in order for this to be possible it is necessary for some gross, basic computations to be made. These separate out a particular pattern from the complex of physical stimuli impinging on the sensory end-organ, a complex representing the interaction of many patterns. The results of this crude analysis permit the system to readjust itself so as to process the details of this pattern. He uses the term "preattentive processes" to refer to these preliminary operations. Solley and Murphy also link the process of selective attention directly to a preceding stage involving the development of "perceptual expectancies." It is logical to assume that a primary need, such as self-preservation, tension reduction, or social adaptation creates the need to perceive a certain pattern and leads to the selection of criteria based on the prediction or expectancy that some patterns will be more relevant than others. This may then lead to the weighting process which facilitates the processing of relevant features while inhibiting, to varying degrees, less relevant features.

The end result of selective attention is the establishment of the *figure-ground relationship*. This phenomenon occurs when a particular pattern stands out from the complex, forcing other pattern groupings to be perceived as background. It is apparent in our ability to listen to one speaker among a group, or to pay attention to one type of instrument while the whole orchestra is playing. The emphasis, however, is on *control* of the figure-ground relationship, rather than upon the organism's ability to focus selectively (an innate ability, at least in visual perception) (von Senden, 1960; Lawson, 1967). Thus, control of the figure-ground relationship, the preattentive processes, and the development of perceptual expectancies are all aspects of the same perceptual process. One finding supportive of such a process has been that cortical response to visual and auditory stimuli is enhanced when the subject is instructed to attend to the stimuli (Spong et al., 1965). It may well be that when we "pay attention" to something information is processed differently.

In the processing of speech, perhaps the percept is in most instances arrived at from only relatively crude data. This would underline the importance of our expectancies with regard to linguistic, contextual, and situational constraints. The need to pay attention to individual components thus becomes unnecessary except when the content is unfamiliar or unexpected, when the speech production pattern deviates significantly from our own, as in a foreign dialect, or when there is very little redundancy in the message signal. This idea would also help to account for the fact that we are capable of understanding speech at a rate far greater than the cochlea can separate the acoustic signal into its component parts (Liberman et al., 1967); it would explain, too, why we perceive foreign language speakers as generating speech at seemingly phenomenal rates. We will defer our discussion of auditory attention mechanisms until the next chapter, but it does appear that auditory perception, like other sensori-perceptual processes, involves the checking out of perceptual expectancies. Moreover, it seems that the degree of attention necessary to perform this task successfully will be a function of the accuracy with which those predictions can be made. The accuracy in turn is determined by the amount of structure the individual is able to impose upon the task.

Summary

Our knowledge of the structure of the acoustic signal of speech, particularly its temporal aspects, together with even the limited amount of neurophysiological information available, encourages a view of auditory perception based upon the internal restructuring of the complex pattern, rather than upon a simple matching concept. Our model has therefore evolved into a "search and synthesize" system, in which the features of particular acoustic patterns are categorized. This process was described as one which progressively narrows down the possible perceptions by ruling out unlikely possibilities. In this manner each decision permits the more effective tuning of the advanced stages of processing. We considered possible strategies by which this end may be achieved; we contrasted the serial and parallel strategies, concluding that speech perception, under normal conversational circumstances, is probably achieved through the more rapid parallel strategy.

Our discussion of computer models employing these two strategies was stimulated by the potential such models have for self-evaluation of their internal logic and by their ability to perform what they attempt to explain. We were also interested in them for their capacity to adjust to various stimuli on the basis of the effectiveness of the program in previous situations. This adjustment process, based upon feedback and the use of a continuous computation of future probabilities, is an important characteristic of the human perceptual system. A predictive process is an essential component of perception, since it is necessary for the system to be selective in its search for relevant patterns. We discussed this process of selective attention as a biological tuning of the sensori-perceptual system, achieved by the differential behavior of neuronal groupings. Such selective tuning was considered to provide an "editing" at various levels in the sensori-perceptual pathway, resulting in the front page headlining of some information, while relegating progressively less important information to the inside pages. The point of the analogy is that any informational item is available on command to the higher levels of perception. The command would simply refocus attention as a result of a reordering of probabilities arising from changing needs.

Our task now is to examine the specific nature of speech perception in the light of the background material presented in the past three chapters.

REFERENCES

ADRIAN, E. C., 1947. *The Physical Background of Perception*. Oxford: Clarendon Press, pp. 76–78.

ARBIB, M. A., 1964. *Brains, Machines and Mathematics*. New York: McGraw Hill.

BERRY, MILDRED F., 1969. *Language Disorders of Children*. Appleton-Century-Crofts: New York.

BRAZIER, M. A. B., December, 1964. The electrical activity of the nervous system. *Science*, 146, 1427.

FANT, G., 1967. "Auditory Patterns of Speech," in *Models for the Perception of Speech and Visual Form*, ed. W. Wathen-Dunn. Cambridge, Mass.: The M.I.T. Press, pp. 111–25.

GIBSON, J. J., 1966. *The Senses Considered as Perceptual Systems*. Boston: Houghton-Mifflin Co., p. 49.

KOLERS, P. A., 1967. "Comments on the Session on Visual Recognition," in *Models for the Perception of Speech and Visual Form*, ed. W. Wathen-Dunn. Cambridge, Mass.: The M.I.T. Press.

LADEFOGED, P. The perception of speech. In *Nat. Phys. Lab. Symp.* 10, 1959. Mechanization of thought processes. I: London, HMSO, 1959, pp. 397–409.

LAWSON, CHESTER A., 1967. *Brain Mechanisms and Human Learning*. The International Series in the Behavioral Sciences. Boston: Houghton-Mifflin Co.

LIBERMAN, A. M., P. C. DELATTRE, and F. S. COOPER, 1952. The role of selected stimulus variables in the perception of the unvoiced-stop consonants. *Amer. J. Psychol.*, 65, 497–516.

LIBERMAN, A. M., F. S. COOPER, D. P. SHANKWEILER, AND M. STUDDERT-KENNEDY, 1967. Perception of the speech code. *Psychol. Rev.*, 74, 6, 432.

LICKLIDER, J. C. R., 1952. On the process of speech perception. *J. Acoust. Soc. Amer.*, 24, 590–94.

MILLER, G. A. AND S. ISARD, 1963. Some perceptual consequences of linguistic rules. *J. Verb. Learn. Verb. Behav.*, 2, 212–24.

NEISSER, ULRIC, 1967. *Cognitive Psychology*. Century Psychology Series. New York: Appleton-Century-Crofts.

NEISSER, ULRIC, 1963. Decision time without reaction time: experiments in visual scanning. *Amer. J. Psychol.*, 76, 376–85.

SELFRIDGE, O. G., 1959. Pandemonium: a paradigm for learning, Proc NPL Symp 1958, No. 10, *Mechanization of Thought Processes* (Her Majesty's Stationery Office, London, 1959), pp. 513–526.

SENDEN, MARIUS VON, 1960. *Space and Sight. The perception of space and shape in the congenitally blind before and after operation*. London: Methuen and Co., Ltd.

SOLLEY, C. M. AND P. MURPHY, 1960. *Development of the Perceptual World*. New York: Basic Books, Inc.

SPONG, H., R. HAIDER, AND D. B. LINDSLEY, 1965. Selective attentiveness and cortical evoked responses to visual and auditory stimuli. *Science*, **48**, 397.

STERN, W., 1938. *General Psychology*. New York: Macmillan.

STEVENS, K. AND M. HALLE, 1967. "Remarks on Analysis-by-Synthesis and Distinctive Features," in *Models for the Perception of Speech and Visual Form*, ed. W. Wathen-Dunn. Cambridge, Mass.: the M.I.T. Press, pp. 88–102.

UTTLEY, A. M., 1959. Conditional probability computing in the nervous system. Proc NPL Symp 1958, No. 10, *Mechanization of Thought Processes*. London: Her Majesty's Stationery Office, p. 121.

UTTLEY, A. M., 1966. The transmission of information and the effects of local feedback in theoretical and neural networks. *Brain Research,* **2,** 21–50.

5

Theories
of Speech Perception

In the previous chapters, we considered the nature and production of the speech signal. We sought to become somewhat familiar with the nature and operation of the auditory system, which receives and processes the information derived from that signal. We also examined some of the basic characteristics of pattern processing as performed by perceptual systems. Our major aim in this chapter is to examine possible ways in which an idea, existing in one person, can be identified by a second person. I have carefully avoided using such terms as "the manner in which thoughts and ideas are exchanged" or "how messages are sent between individuals," for these are misleading. The problem we face is how a speaker controls which thoughts his listener will have. It is a matter of evoking equivalent ideas rather than of sending or exchanging them.

Reconsider for a moment *what* the acoustic signal actually is. The sound wave consists quite simply of variations of air pressure impinging upon the eardrum. At any given instant in time the information derived from the acoustic signal is expressed in terms of the amount of energy present relative to a reference level and the manner of distribution of that energy across a range of frequencies. If, instead of taking a single reading, we stretch our examination of the physical stimulus over a period of time, we begin to observe something different: the *acoustic event*. This comprises not simply a continuous series of pressure changes, but a series of relation-ships of pressure values and energy distribution. We can study the relative as well as the absolute values. We can record not only the changes, but also the nature of these changes. The acoustic stimulus is, therefore, a series of time-related events. The complete record of events constitutes the source of information from which we will assign probable perceptual values. *The acoustic signals of speech do not in themselves convey messages:* they do convey *information*, from which messages equivalent to the speaker's are hopefully reconstructed by the listener. Over time vibratory events, both speech sounds and nonspeech sounds, impose a structure upon the move-ment of the air molecules they disturb. Each vibratory event patterns the sound wave in a manner peculiar to its vibratory structure. The sound wave is imprinted with this information which is, therefore, indicative of the event. The sound wave thus becomes referential.

On reaching the auditory system the pattern changes the status of the organ at the periphery, constraining and structuring its behavior. Because of the tuning effects of the efferent system, the responsiveness of the sys-tem to certain patterns will be influenced by its particular "set" at a given time. Nevertheless, to a varying extent an acoustic event which occurs within the sensitivity of the organism will induce change in the neural status of that organism. The induced change will be patterned in a manner *equiv-alent to*, but not identical with, or even similar to, the stimulus pattern. The equivalence of the patterns maintains the continuity of information flow. What we receive, therefore, is information concerning how the acoustic event has influenced or structured our system (relative to its own particular posture at the time of the event) and the manner in which it was processed by our system.

Thus, auditory perception of speech is a process of interpreting the in-structions imprinted on the acoustic wave by the speaker over a time span, or as Berry (1969, p. 59) states, "Auditory perception of speech per se deals mainly with the temporal management of information from the input."

It is very important to recognize that speech is a continuous, unseg-mented event. When we examined the articulatory production of speech, we saw that the organs of speech glide from one target position to the next, generating transitional information in the process. There are no distinct jumps from the articulatory posture associated with one speech-sound value (phoneme) to the next, even though we appear to perceive phonemes as existing in distinct categories with clearly defined perceptual boundaries. Also, the characteristics of the acoustic stimulus for any given phoneme are considerably influenced by its neighbors, i.e., its phonetic context. The rapidity with which we produce speech results in the phenomenon of *co-articulation*, which is the overlapping of the articulatory constituents of one sound with the next. Co-articulation can modify the manner of production and, therefore, the acoustic pattern of a sound preceding or succeeding a given phoneme. An example of the latter would be the continuation the voiced /d/ into the unvoiced phoneme /s/ as in the word "cards" where (when spoken rapidly) the /s/ may be partially voiced; e.g., "The cards are on the table." The effects of co-articulation may also project well ahead of the sound being articulated. Daniloff states:

> . . . what is observed is that when given the chance, articulator move-
> ments are often initiated far in advance of the sound in which they
> are crucial (anticipatory positioning). This is the so-called right-to-left
> co-articulation. (1973, p. 205)

It is apparent from examination of spectrograms of connected speech that there is a discrepancy between what the acoustic waveform is seen to be doing, and what we perceive auditorily. To quote Daniloff again:

> Articulation processes run together and overlap the phoneme-sized
> units so that the phonemes are allophonically varied and changeable.
> From this shifting continuous stream, listeners can decode speech and
> extract the phoneme units, even though it is difficult to observe inde-
> pendent, non-overlapping sounds in the speech acoustic wave (Liber-
> man et al., 1967). It is this seeming divergence between production
> and perception (Kozhevnikov and Christovich, 1965) and between
> larger, perhaps syllable sized production units and smaller perceptual
> ones which puzzles students of articulation. (1973, p. 208)

It has been pointed out that the production of a speech sound requires a degree of *invariance* if the sound is to be identifiable. The difficulty is to determine the nature of the invariance. Since there is a causal relationship between the manner in which a speech sound is produced and the resulting acoustic wave, we can consider the perception of any sound in terms of either (a) *the manner of articulation used in its production,* or (b) in terms of *the resultant acoustic event.* Therefore, the problem hinges on how linguistic values are recovered from the speech wave and on the nature of the basic unit of speech.

In the search for an explanation of how linguistic value is determined from a speech signal, two approaches have been taken. The terms "active" and "passive," originally used by McKay (1956) describe these two ap- proaches. The passive system is envisaged as a filtering system functioning to identify and combine information so as to restructure the pattern. Today this designation has been modified to yield a more flexible, tuneable system. Active models, by contrast, are viewed by McKay as "comparator" systems, in which an input pattern is compared to an internally generated pattern. Active theories necessitate a mediational process. Thus the so-called active and passive theories might be better termed *non-mediated* (passive) and *mediated* (active). The major theorists in these two schools are summarized in Table 5.1.

Table 5–1
Abstract of Major Exponents of Passive and Active Theories of Speech Perception

PASSIVE THEORISTS
(Non-mediated, direct decoding)

Selfridge

"Pandemonium Model." Complex signal interrogated by "demons" (computational, cognitive, decision) to determine the presence or absence of the features they represent. System is tuneable by weighting relationship patterns between computational and cognitive levels.

Uttley

Neurological model based on mathematical classification of the CNS. Binary analysis leading to identification of a unique pattern of firings. Four requirements must be met:

(1) Input channels identifiable as active/inactive
(2) Multiple combination of inputs
(3) Indicator unit for each combination denoting *active* status of each component thus defining the pattern
(4) Storage capacity

Fant

Source (voice)/Filter (articulators and resonators) Model which uses distinctive-feature theory. Source characteristics:

(a) silence   (c) voice and noise
(b) voice   (d) noise.

Acoustic features mapped onto neurological features of auditory system. This results in internal auditory pattern representations. Acknowledges parallel encoding at phonemic level and need for syllabic processing.

Hughes and Hemdal

Computer-based Model. Acoustic signal analyzed in binary form according to distinctive features for categories of sound. Each decision determines the routing of subsequent interrogation. It does this by computing the probabilities of occurrence from the data already available. System is self-adjusting (tuneable) to allow for inter-speaker variation (normalization).

Abbs and Sussman

Neurological model. Feature detection based on auditory decoding. Features represented as specialized neural configurations. Parallel processing of information of transitional cues. Efferent control permits neural tuning through selective lateral inhibition. Rapid sorting of speech sounds into categories according to distinctive features.

ACTIVE THEORISTS
(Mediated, indirect coding)
Liberman et al. (Haskins Lab Group)

Postulate a motor theory based on the mediative role of the neuromotor rules

Table 5-1 (continued)

of speech production. Propose that the model operates from phonemic to semantic levels of linguistic processing. Speech is perceived by reference to manner and place of production of the speech wave. Speech production and perception seen as unique process involving

(a) a special function
(b) a distinctive form
(c) use of a special key to encode/ decode
(d) a special perceptual mode. Sees speech as hierarchically structured at all levels. Parallel encoding and decoding. Relates parallel encoding to co-articulatory function and syntax. Segmental analysis is recognized. Speech is perceived by running the production process backward.

Stevens and Halle

Analysis-by-synthesis model based upon hypothesis testing. This involves perceiving speech by comparing incoming signal (acoustically and linguistically) to internally generated patterns. These internal representations are generated according to rules selected on probabilities based on earlier data. Concur with Liberman that generative rules are highly similar to those used in production. Abstract representation of speech event fundamental to production and perception. Identification of appropriate generative rules permits invariance to be maintained through computing of error correction (normalization).

Passive (Non-Mediated) Theories

It might be reasonable to assume that the acoustic pattern produced by the articulatory resonator system of the speech mechanism is unique for each speech sound. Until recently most of the research on speech perception sought to determine the nature of the supposed "direct relationship" between the acoustic signal and the phoneme value. However, recent evidence does not support the existence of such a direct relationship. Our previous discussion of the acoustic characteristics of speech sounds revealed that the frequency components of the sound wave, generated as a function of the phoneme value to be communicated, vary from one production to the next. This is true both within a given speaker's utterances and between different speakers. The acoustic pattern is influenced both by the pitch of the voice and by the effects of phonetic context. We saw how the effects of co-articulation result in changes which extend even beyond immediate neighbors. Later we will examine further evidence which refutes the concept of speech as a series of discrete units. Yet, whether it is to be found in the acoustic pattern, in the articulatory pattern, or in both, the criterion of distinctiveness must be met in some way if consistency of perception is to remain a premise.

Jakobson and Halle (1956) and later Jakobson, Fant, and Halle (1963) have suggested a set of distinctive features which combine both acoustic and articulatory characteristics. The features they propose arise from the limited range of patterns which man's articulatory resonant system is capable of imposing on the glottal tone. The proposed set of features shown have been tabulated by Table 5.2, p. 102.

These features constitute a binary code system. We considered the nature of such systems on pages 80–82. Involved was a decision-making situation in which one of two criteria were identified. For example:

Is sound energy present?   yes/no

Is voicing present?   yes/no

Is nasality present?   yes/no

According to this model, the identification of a combination of yes/no answers to different questions will constitute a pattern whose features will be distinctive of a certain phonemic value for a given language. The pattern for each phoneme is quite distinctive. A major advantage of the feature system is that a single feature has the capacity of discriminating between several pairs of phonemes.

The answers to each of the interrogations must in some form exist in, or be derivable from, the acoustic event. Our discussion of the acoustic speech signal in Chapter 2 provided us with some clues about how infor-

Table 5–2
The Distinctive Features of English

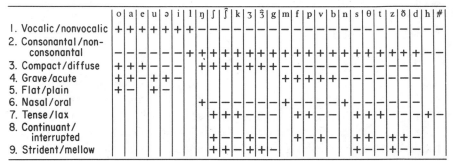

	o	a	e	u	ə	i	l	ŋ	ʃ	ʃ̂	k	ʒ	ʒ̂	g	m	f	p	v	b	n	s	θ	t	z	ð	d	h	#
1. Vocalic/nonvocalic	+	+	+	+	+	+	+	—	—	—	—	—	—	—	—	—	—	—	—	—	—	—	—	—	—	—	—	—
2. Consonantal/non-consonantal	—	—	—	—	—	—	+	+	+	+	+	+	+	+	+	+	+	+	+	+	+	+	+	+	+	+	—	—
3. Compact/diffuse	+	+	+	—	—	—		+	+	+	+	+	+	+														
4. Grave/acute	+	+	—	+	+	—									+	+	+	+	+									
5. Flat/plain	+	—		+	—																							
6. Nasal/oral								+	—	—	—	—	—	—	+	—	—	—	—	+								
7. Tense/lax									+	+	+	—	—	—		+	+	—			+	+	+	—	—	—	+	—
8. Continuant/interrupted									+	—	—	+	—	—		+	—	+	—		+	+	—	+	+	—		
9. Strident/mellow									+	+	—	+	+	—		+	—	+	—		+	—	—	+	—	—		

Key to phonemic transcription: /o/–pot, /a/–pat; /e/–pet, /u/–put, /ə/–putt, /i/–pit, /l/–lull, /ŋ/–lung, /ʃ/–ship, /ʃ̂/–chip, /k/–kip, /ʒ/–azure, /ʒ̂/–juice, /g/–goose, /m/–mill, /f/–fill, /p/–pill, /v/–vim, /b/–bill, /n/–nil, /s/–sill, /θ/–thill, /t/–till, /z/–zip, /ð/–this, /d/–dill, /h/–hill, /#/–_ill.

(From *Preliminaries to Speech Analysis* by R. Jakobson and M. Halle, 1956.)

mation exists in the sound stream. We saw that voicing generates a glottal tone with energy centered between approximately 133 Hz for male speakers and 199 Hz for children's voices. Thus the differentiation between voiced and unvoiced sounds can be made by determining whether there is energy below 300 Hz. The identification of the sound as one of the voiced group of English speech sounds will structure the type of processing to which the input data is then subjected. Previously we saw that the glottal tone causes the air in the resonant cavities of the speech system to vibrate in a manner characteristic of their posture. It was explained that this results in a rearrangement of the energy in the glottal tone. Certain speech sounds have heavy energy concentrations in certain frequency bands. Thus, for example, if the voiced sound is found to have a concentration of energy in the frequency range above 3000 Hz, the sound must be one of the group /ð/v/z/ʒ/. The vowel sounds can be identified by the fact that the relative intensity of vowels is greater than that of consonants. They can also be differentiated partly on the basis of the relationship of the first two formants. Some consonants such as /m/n/l/r/w/j/ are characterized by a complex structure very similar to vowels. However, the intensity of these sounds is less than that of vowels, and /m/n/ have the added nasal component. The stop consonants /p/b/, /k/g/, and /t/d/ may be identified as a group by the period of silence which precedes them. The presence or absence of voicing will permit further discrimination between /p/t/k/ and /b/d/g/ as groups.

A series of binary questions progressively narrows down the speech-sound category and thus the phonemic value to be associated with the input stimulus.

The "Pandemonium" model, considered in the previous chapter, accommodates such a non-mediated theory of speech analysis. Each demon represents filter units sensitive to certain relationships of frequency, intensity, and duration present in the speech wave that is processed by the cochlea. The data demons receive the crudely analyzed cochlea information, record it, and then pass it to the bank of computational demons arranged in parallel. Each of these establishes the presence or absence of the articulatory/acoustic features described, for example, by Jakobson. Whether a given variable-filter unit (demon) fires or not is dependent upon the presence or absence in the acoustic waves of the particular feature to which it is sensitive. The output of these distinctive-feature analyzers is then fed to the cognitive level where more complex, less specific units are activated to varying degrees.

As was pointed out earlier, the feature characteristics of phonemes overlap to a considerable extent. For example, one single feature is responsible for the differentiation of /p/ and /b/ or /v/ and /b/. Thus some of the criteria satisfy several of the cognitive demons, but only the one whose criteria are completely satisfied causes the decision demon to identify the appropriate phoneme value.

The Acoustic Theory

Fant (1960, 1962, 1967) has also sought to explain both the production and perception of speech by utilizing the concept of distinctive features, which he developed with Jakobson and Halle. He traces the origins of the speech wave pattern to "the response of the vocal tract filter systems to one or more sound sources." Using this concept it becomes possible to specify the speech wave in terms of its source and filter characteristics.

The source or voice characteristics can be specified as:

(1) silence   (3) voice + noise
(2) voice   (4) noise (i.e., non-voiced speech sound).

The vocal stimulus or the noise source is then filtered by the articulatory resonator system (Chapter 2, p. 25), which arranges the formant structure in a manner characteristic of the articulatory-resonant posture peculiar to the speech sound being generated within a phonetic context. A series of such postures, with their associated linking transitions, results in the speech wave. Fant maintains that the boundaries of the successive sound units can be fairly well defined in terms of the source and the filter. He recognizes that these boundaries are phonetic (acoustic) and not phonemic (linguistic) in nature, but, he argues, these sound units are the building blocks of speech. It is from these units that the binary information pertaining to distinctive features is extracted. He believes that the acoustic patterns of speech are then mapped onto the neurophysiological structures of the auditory system. In this way the distinctive-feature information characterizing the phonemes of speech is represented in the neural system. Thus the distinctive-feature information "exists" in the articulatory stage of speech production, is imprinted upon the acoustic speech wave, and is then internalized by the listener who draws physiological maps in the auditory system (Fant, 1967, p. 112). These maps then constitute the internal auditory pattern representations.

Fant acknowledges the fact that there is considerable overlapping of phonetic influence of sounds. This overlapping is compatible with the theory of distinctive-feature analysis. He states:

Distinctive feature analysis applied to speech does not require an initial stage of segmentation in terms of sharply time limited portions of the speech wave (Fant 1962). Some features appear and fade out gradually, and the tendency of segmental structuring to be observed in spectrograms is such that one phoneme is often characterized by cues from several adjacent segments and that one segment may carry information on the identity of two or more successive phonemes. (1967, p. 118)

We shall see that the overlapping of information may be a major factor in phoneme identification. The coexistence of distinctive-feature information pertaining to adjacent phones, as well as to the phone under study, deprives the speech sound of an independent identity. As Fant explains:

If we want to describe one of these features, it is not sufficient to cut out the segment traditionally ascribed to the consonant and observe it. We can, of course, observe some general differences in terms of such segments, but if we gate out a single one of them, e.g., the $/ l /$ segment defined by oral central closure and lateral opening, and listen to it, we will not hear an $/ l /$ but a vowel. (p. 119)

Fant then makes the important point that:

It is the contrast with adjacent sound segments that accounts for the specific $/ l /$ quality. (p. 119)

We shall encounter this concept of the phoneme being cued by its neighbors in our discussion of the motor theory.

Fant maintains, therefore, that for most speech sounds the auditory information at the periphery is directly (passively) encoded into distinctive auditory features. He suggests, however, that for some phonemes and even for some syllables or words, recognition might be possible immediately without prior analysis into distinctive features. In general, though, the acoustic theory is dependent upon the extraction from the acoustic signal of those physical characteristics which represent the distinctive features.

Hughes and Hemdal (1965, 1967) have reported on a computer program based on such a non-mediated model. The physical attributes of the acoustic signal to be processed by the computer were those representing the distinctive features proposed by Jakobson and Jakobson, Fant, and Halle. The researcher's concern was to determine whether a computer, programmed to represent the non-mediated, theoretical model outlined above, could in fact perform the primary recognition tasks of the auditory system: the identification of phonemes. The results of the experiment were impressive, since it was shown that for 227 nonsense words the computer correctly identified 90 percent of the phonemes. This is equivalent to human performance. The computer was programmed to identify ten cardinal vowels, nine diphthongs, and most of the consonants. Because of the variation of the acoustic event generated between different speakers uttering the same speech sound, it was necessary for the computer to be able to tune itself to the pattern of speech characteristic of each speaker. Fant (1956) and Ladefoged and Broadbent (1957) have suggested that such a tuning process may play an essential role in the processing of speech by the human auditory system. Apparently it "normalizes," or adjusts, the incoming pattern to the standards of the listener's speech and thus compensates for disparities between the speaker's speech model and the listener's (see p. 27).

The Hughes and Hemdal (1967) program analyzes each segment of the acoustic signal as shown in Figure 5.1. The answer to each interrogation progressively rules out groups of sounds and presets the system to continue analysis commensurate with the probabilities already derived.

Our earlier examination of the auditory system suggested that some sort of ongoing tuning of the system on the basis of information already processed is almost certainly involved in auditory perception. The rich supply of interconnecting short and long axon neurons present at each level and between levels and running both upward and downward, as well as the efferent system of fibers, lends support to the idea of the lower systems exerting control on the mode of processing carried out at higher levels. As in the computer model, the analysis would be progressively refined by the nature of the data already derived.

Abbs and Sussman (1971) have given considerable thought to how such neurological processing takes place. They have developed a neurological theory of speech perception compatible with the concepts we have already considered.

Neurological Theories

As Corcoran (1971) has explained, passive processing of patterns involves a two stage process: "a) *analysis* into their parts, and b) the *resynthesis* of the processed parts back into a neurological representation of the entire stimulus."

A neurological theory must postulate the existence of specialized nerve cells, or groups of cells. These must be sensitive to and capable of analyzing information. During the past fifteen years or so, physiologists have begun to provide evidence that specialized receptors exist in more than one sensori-perceptual system. Investigation of the neurophysiological structure of the visual system of animals has demonstrated that patterns of cell groupings exist and are selectively responsive to specific features of the

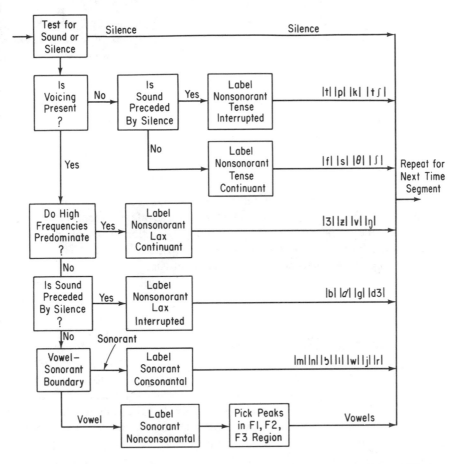

Figure 5.1   Independent time segmentation classification. (Hemdal and Hughes
1967, p. 448))

visual stimulus. These patterns include the direction, motion, curvature,
and rate of change in brightness of a contour (Hubel and Wiesel, 1963;
Barlow and Hill, 1963; and Lettvin, Maturana, McCulloch and Pitts
1959). Similar sensitivity has been demonstrated to exist among some cells
of the auditory system (Evans and Whitfield, 1964; Whitfield and Evans,
1965; Goldstein, Hall, and Butterfield, 1968; Hall and Goldstein, 1968;
Frischkopf and Goldstein, 1963). Abbs and Sussman (1971) have provided a review of most of this evidence which they use as a basis for the
neurologically oriented, non-mediated theory of speech perception. They
propose a feature-detector theory "which does not depend on a particular
distinctive system, but rather concerns itself with the process of auditory
decoding of the acoustic speech signal which results in phonetic identifica-

tion" (p. 23). They define feature detectors as "organizational configurations of the sensory nervous system that are highly sensitive to certain parameters of complex stimuli" (p. 24). Unlike acoustic filters, which are capable only of sampling the energy within a narrow band, feature detectors respond simultaneously to multiple-stimulus characteristics, most of which arise from spatial-temporal changes in the acoustic wave. Abbs and Sussman cite evidence from visual neurophysiology (Maturana and Frenk, 1963) to indicate that feature detectors demonstrate a range of tolerance for stimulus parameter variation. Such tolerance would be important to a system which deals with a stimulus input variation similar to speech signal variations for a given speech sound uttered on repeated occasions. The boundaries of such tolerance might account for the perceptual boundaries we experience between phonemes. Drawing on evidence from the same researchers, they point out that at least in pigeons efferent motor impulses enhance the sensitivity of certain neuronal groups. They state: "This process acts to enhance differences and reduce confusion at the periphery and at higher stations by the inhibition of one nerve cell by another without specific central control" (p. 25).

Abbs and Sussman (1971) suggest that data on neuron patterns in the visual system may have implications for the auditory system. They point out that some of the neuronal cell groupings of the visual system are sensitive only to complex stimulus patterns. Such patterns are analogous to the complex patterns of speech. The sensitivity of these cell groups (the criteria which must be satisfied before they will fire) is quite specifically defined in terms of a range of stimulus characteristics. Finally, they suggest that such neuronal patterns, tuned by the efferent system, may explain how invariant perception of phonemes is possible in the presence of a variable acoustic signal.

The presence in the auditory system of neuronal groups sensitive to complex stimuli has been documented by several researchers (see Chapter 3, p. 74). Abbs and Sussman argue that "A complex spatial configuration of receptor cells, maximally sensitive to certain physical parameters of the auditory stimulus, can be postulated. Such cells may offer an explanation for the neural decoding process involved in speech perception" (p. 28). Also ". . . spatial configurations of receptor cells located in the inner ear can be especially tuned to respond to formant patterns and most notably formant transitions" (p. 28).

We discussed the importance of formant transitions when we examined the acoustic signal. Reference was made to Liberman's (1957) claim that the importance of the transitions cannot be exaggerated. The neurological theory advanced by Abbs and Sussman makes it possible to explain theoretically how the transition changes, coded into the spatiotemporal aspects of the acoustic wave, may be detected by the auditory system. The complex waveform, which reaches the cochlear duct as a hydraulic disturbance, results in a pattern of stimulation along its length and permits the recording of the spatial relationship of the frequency and intensity information contained in the waveform. Since the energy wave is a continuous event, the changing pattern of these relationships may be temporally detected. This spatiotemporal analysis is further carried out by the time pattern of firing of spatial patterns of neurons. The transition cues in the acoustic wave may be internalized and analyzed in this manner by the auditory system. In order for triggering to occur, each criterion for firing would have to be met within the limits of tolerance of the neural cells in the complex. Such an arrangement, augmented by the tuning action of lateral inhibition, could explain how the system identifies phonemes which differ by only one feature. The work of Miller and Nicely (1955) on perceptual confusions lends support to this concept. The boundaries of tolerance for firings might account for the system's perceptual tolerance of variation in the acoustic equivalent of a given phoneme.

In their article, Abbs and Sussman present both psychophysical and developmental evidence in support of their theory. They point out that, if correct, their theory would require that speech stimuli would be processed differently from nonspeech stimuli of equal complexity. The question is, does the brain in fact treat these two kinds of stimuli differently? A similar question was asked about the visual system by Corcoran and Rouse (1970), who were interested in whether handwriting and printing are processed by the same system. The results of their experiment indicated that the two types of stimuli are processed differently.

A number of researchers (Broadbent and Gregory, 1964; Chaney and Webster, 1965; Kimura, 1961, 1967, 1973; Milner, 1962; Studdert-Kennedy and Shankweiler, 1970; Mattingly, Liberman, Syrdal, and Halwes, 1971) have provided evidence that speech sounds and nonspeech sounds other than rhythmical features are processed in opposite hemispheres of the brain; speech mainly in the left, nonspeech in the right. Of course, the fact that the two types of materials are processed by different hemispheres is not itself conclusive evidence that they are necessarily processed differently.

Abbs and Sussman (1971) cite an experiment by Warren, Obusek, Farmer, and Warren (1969) in support of two different modes of processing. The experiment demonstrated that short-duration (70–80 msec.) speech sounds of normal conversation are obviously perceived in correct sequential order. But when presented with a task of perceiving the temporal order of three successive nonspeech sounds, subjects were unable to remember the order of presentation, even when given a much longer exposure time

(200 msec.). This temporal processing difficulty failed to disappear even after training, yet when spoken digits were substituted for nonspeech sounds the task was performed with ease.

Studies bearing on the categorical perception of consonant phonemes are further sources of evidence to suggest that there exists a distinctive mode for the processing of speech. Psychophysical experiments have demonstrated that when nonspeech stimuli are varied over a continuum, the listener hears the continuous variation as a continuum. Subjects can be taught to identify arbitrarily determined boundaries along the continuum and thus to recognize discrete categories, though often not without difficulty. The auditory perceptual mode for nonspeech signals involves, therefore, *continuous perception.* As Studdert-Kennedy, Liberman, Harns and Cooper state:

> Continuously perceived stimuli . . . are perceived relationally, and even if placed in classes (categories), may be perceived as different. To discriminate between them is to detect a difference, whether or not they are identified as belonging in the same class. (1970, p. 236)

Research on the perception of speech sounds indicates that "For certain consonant distinctions it has been found that the mode of perception is in fact nearly categorical" (A. M. Liberman, Cooper, Shankweiler, and Studdert-Kennedy, 1967, p. 442). *Categorical perception* involves the absolute perception of the stimuli. Instead of perceiving the progressive changes in a continuously varying stimulus, in this case the acoustic speech stimulus, listeners report the perception of jumps from one perceptual category to another. Within the boundaries of these perceptual categories little change is noticed, even though the acoustic stimulus, in fact, is continuously changing. Once a boundary is crossed, the perceived stimulus information is recategorized and attributed a new value.

> In particular we have found that certain speech sounds (stop consonants) tend to be perceived categorically, while certain other speech (vowels) and nonspeech sounds tend to be perceived continuously. (Studdert-Kennedy, Liberman, Harns, and Cooper, 1970, p. 236)

Experimental work by Bastian in cooperation with Eimas and Liberman (1961) showed that the discrimination between the words "slit" and "split" can be made purely on the basis of the duration of an interval of silence which occurs between the friction of /s/ and the vocalized portion which follows. At a given duration-boundary the perception categorically jumps to the other phoneme. Similar examples involving other phoneme categories have been cited in the literature (Liberman, Cooper, Harns, MacNeilage, and Studdert-Kennedy, 1967b, p. 69).

Abbs and Sussman interpret these data to suggest that:

> . . . speech sounds are quickly and accurately sorted into the proper bins according to distinctive acoustic features, for it seems as if the neural system decoding speech has developed a unique and high level detection process specialized for the space-time properties of spoken language. (1971, p. 31)

Finally, the authors of the neurological theory cite developmental evidence which suggests the existence of innate feature-sensing neuron systems stimulated in their development by exposure to spoken language. They consider this exposure to be critical to the establishment of the patterns of neuron arrangement peculiar to the perception of a particular language. The findings of McCaffrey (1967) and Moffitt (1971) are cited. The results of these studies indicate that the auditory system of very young infants discriminates between synthetic speech patterns of certain consonant sounds. Change in heart rate (which decelerates when presented with a stimulus perceived as new) occurred on presentation of a second consonant sound after the infant had become accustomed (habituated) to the first. This indicates that the acoustic features were identified as dissimilar. In fact, differentiation was demonstrated between phonemes when the only identifying factor was the difference in the second formant transition. Similar findings have since been reported by other researchers. These are discussed in Chapter 7. Abbs and Sussman in conclusion state:

> A feature detector model of speech perception can provide a direct explanation of a very intricate transduction-detection phenomenon, changing acoustic energy into coded neural energy at high rates of acoustic input. Furthermore, the inclusion of selective lateral inhibition operating at the levels of the VIIIth nerve and higher provides a mechanism which further outlines an explanation of invariant perception with variant acoustic input. (1971, p. 34)

Abbs and Sussman have presented a convincing argument in favor of a passive system of auditory processing, involving specialized groups of cells capable of detecting the presence or absence of certain distinctive features in the acoustic signal. Their theory is in agreement with the earlier work by Uttley (1959, 1966).

Uttley has worked on developing a theory to explain how the neural system computes information so as to permit classification of stimulus patterns. He lists four requirements to be met in order that the mathematical principle of classification be operational in the nervous system. These requirements as detailed in his 1959 publication are:

1. Each input channel must be always in one of two states—active and inactive.

In the binary system so constituted, the activity or nonactivity of a particular neuron may, therefore, represent either one of any two criteria; for example, voicing or non-voicing, plosion or absence of plosion.

2. The inputs must be combined in as many ways as possible— ideally in all possible ways. (p. 122)

Each combination of connections contributes to a unique neural pattern with its corresponding acoustic pattern. Each would correspond to the "complex spatial configuration of receptor cells" (to which Abbs and Sussman refer), each tuned to respond to a particular formant pattern.

3. There must be a unit for every combination of inputs, which *indicates* if every input of the combination is active. A combination or set of inputs is said to define a *pattern* of activity. (p. 124)

This requirement is critical for the perception of the speech pattern. The spatial pattern of inputs must represent or *synthesize*, for a given instant of time, the intensity/frequency relationship of the acoustic input. This is derived from the analytical data generated at the peripheral level of the cochlea.

The final requirement is that

4. . . . there shall be some way of delaying signals for periods of the order of seconds. (p. 124)

This is essential to the synthesis of the temporal aspects of the serial-ordered signal. The temporal processing permits the identification of pattern information which can only be built up over time. Uttley presents evidence to support his claim that each of these requirements is in fact satisfied by the nervous system. Central to Uttley's model is the concept of *indicator* units, which require more than one input to trigger them. The multiple inputs mean that these units are capable of synthesizing analyzed data, since each indicator unit is *equivalent* to a particular spatial arrangement of peripheral analytical units (Figure 5.2). As in the figure, sixteen

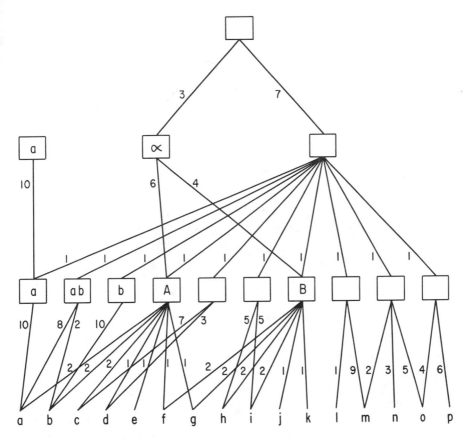

Figure 5.2 Classification by means of identical units, each of which indicates if a fixed number of inputs to it become active, when the number is ten. (Uttley, 1959, p. 125)

peripheral units (a – p) fire into ten indicator units: a, ab, b, A, etc. Each represents a discrete pattern determined by the pattern of inputs from the periphery. Uttley suggests, therefore, that "each neuron distinguishes a particular pattern of activity by virtue of its connections" (1959, p. 128). He stresses, as we have, that the system is dealing with the *patterns of activity aroused within it by the external stimulus*, not "the limitless patterns of the external environment which can arouse the internal patterns." This concurs with our earlier interpretation that what we are processing is information concerning how our system is functioning, or being modified, as a result of the peripheral stimulus. As Corcoran (1971, p. 113) succinctly states, "The response to a pattern is itself a pattern of efferent impulses to a group of effectors." The system is designed to reduce the redundancy present in the input signal. The multidimensional stimulus, which at lower levels excites a number of analyzers, is represented by fewer and fewer neurons as the complexity of the pattern is progressively synthesized. This occurs because at the higher levels the indicators, which represent subpatterns of increasing complexity, are themselves triggered by coded multidimensional information from which the overall pattern is finally reconstructed. The interpretation of data in terms of how it modifies our system would make it possible to meet the criterion of invariance. A system such as the auditory perceptual system is required to derive consistency from variable stimuli, and in this manner the effects of speaker variation and of co-articulation can be accounted for.

The temporal information essential to the reconstruction of the pattern can only be derived by a short-term delay of earlier bits of information, so that they can be combined with later bits before being presented to the indicator units.

It has been shown that for temporal discrimination, each binary signal must pass through a series of delays, and that the output of each delay must provide an input to a classification system. (Uttley, 1959, p. 131)

Such delaying mechanisms may well be achieved through the gating mechanism, *lateral inhibition*, which we discussed in Chapter 3 (p. 72), and which Abbs and Sussman (1971) suggest as a contributing factor in the process of perceptual invariance (see quote, Chapter 5, p. 111).

The passive or non-mediated theories of speech perception assert, therefore, that the process of auditory perception involves the direct structuring of active neural patterns by the pattern information content (distinctive features) of the acoustic stimulus. It is not suggested that permanent arrangements of neuron patterns exist for each sound value, but that arrangements are temporarily structured by specific combinations of acoustic energy. The perception of a particular speech sound would, therefore, be dependent upon the detection in the acoustic signal of those parameters peculiar to the criteria for that speech sound value for that individual. Fant (1967) refers to the "physiological mappings of speech wave patterns in the auditory system." Given fairly broad criteria, it is possible for such a system to tolerate the natural variations in speech sound production by a speaker and between speakers. The tolerance can be further enhanced by postulating a "normalizing" device which adjusts the listener's system to compensate for the discrepancy between the encoding system of the speaker and the decoding system of the listener. Corcoran in summarizing his discussion of the passive theory suggests that a workable model would have to incorporate the following stages:

(a) the extraction of acoustic features from the stimulus.

(b) the mapping of the outputs from the acoustic feature analyzers onto indicator units whose outputs represent the articulatory features of the utterance.

(c) the mapping of the outputs of the articulatory indicators onto phoneme indicators which respond to patterns of articulatory features with outputs representing phonemes.

(d) the mapping of outputs of the phoneme indicators onto morpheme indicators.

(e) the presentation to the word units of the outputs of the morpheme indicators. (1971, p. 153 – 154)

These ideas are not supported by the advocates of active (mediated) theories.

Active (Mediated) Theories

Those who propose an active concept of the speech-processing system offer evidence to refute the direct passive relationship between the acoustic signal and the perceived linguistic value. Liberman and his colleagues (Liberman, 1957; Liberman, Cooper, Shankweiler, and Studdert-Kennedy, 1967; Liberman, Cooper, Harris, MacNeilage and Studdert-Kennedy, 1967b), at Haskins Laboratories in New York, have conducted experiments designed to examine the process involved in the identification of phonemes from the acoustic stimulus. The evidence which they offer for the rejection of a passive model includes the following:

1. That the length of the phoneme segments at the rates at which we are able to process them, 400 words per minute, or 30 phonemes per second (Orr, Friedman, and Williams, 1965), would average only 50 msec. The number of short-duration sounds which the ear can identify at this rate is considerably less than the forty which constitute the phonemes of English (Nye, 1962). Even at the rate of normal conversation (approximately 15 phonemes per second), if processed as a string of separate acoustic events the phonemes would sound like a buzz.

2. That the perception of a given phoneme is evoked, in different phonetic contexts, by markedly different acoustic patterns. We hear as identical, sound patterns which are in fact physically different.

3. That experimental data indicate that it is all but impossible to cut a tape-recorded utterance into units representing phonemes. When such space/time tape sections are cut and respliced to compose a new word from the individual sections, the identity of the phoneme components is lost: "the speech signal typically does not contain segments corresponding to the discrete and commutable phonemes . . . we cannot cut either the /di/ or the /du/ pattern in such a way as to obtain some piece that will produce /d/ alone. If we cut progressively into the syllable from the righthand end we hear /d/ plus a vowel or a nonspeech sound; at no point will we hear only /d/." (Liberman et al., 1967, p. 436)

4. That the perception of a signal speech sound may be evoked by very different acoustic stimuli, or in some instances in the complete absence of acoustic energy.

These observations, it is argued, all point to the probability that speech perception involves the use of a special code. Instead of direct triggering of neuron groups representing phonemes, Liberman et al. have suggested that the perception of many sounds is mediated by the articulatory process of speech production. "It seems unparsimonious," the authors argue, "to assume that the speaker employs two entirely separate processes of equal status, one for encoding language, the other for decoding it. A simpler assumption is that there is only one process with appropriate linkages between sensory and motor components" (1967, p. 452).

The exponents of active theories thus interpret the experimental evidence on the operation of the human auditory perceptual system to mean that a direct, passive process would inadequately explain the perception of speech. Most of these theorists have concentrated their research on the perception of the speech signal at the phoneme level. It is important to recognize, however, that the system they propose for the processing of phonemes is also intended to accommodate perceptual processing at linguistic levels up to the attribution of meaning. Liberman made this point even in his early work on speech perception.

We might note here that such a system would appear, within limits, to apply at the higher linguistic levels too, where meaning, for example, enters to complicate the psychological picture. Thus, we all know how morphemic elements are entered into various combinations to create a variety of words, each element having a particular identity or meaning which it retains regardless of the combination in which it occurs. Indeed, we may suppose that combining independently variable stimulus elements is a workable basis for perception in the language area. . . . (1957, p. 121)

and again:

We would suggest . . . (that) . . . the operations that occur in the speech decoder—including, in particular, the interdependence of perceptual and productive processes—may be in some sense similar to those that take place at other levels of grammar. If so there would be a special compatibility between the perception of speech sounds and the comprehension of language at higher stages. (Liberman, et al., 1967, p. 456)

The Motor Theory

One of the major active theories of speech perception, the motor theory, has been developed largely by Liberman and his colleagues. This theory is based upon the hypothesis that speech at the level of the phoneme, and probably above, is perceived by reference to the place and manner of production of the acoustic signal. We hear by reference to those neural instructions which we would give to our own articulators in order to effect a signal similar to the one received.

Liberman (1972) has stated that speech perception constitutes a unique process. He has offered four reasons for this opinion:

1. The speech code involves a special function
2. It exists in a distinctive form
3. It is unlocked by a special key
4. It is perceived in a special mode.

The speech code involves a special function: Liberman (1970) and Liberman, Mattingly and Turvey (1972) have suggested that the basis of speech perception probably lies in the grammatical structure of the language. They believe this structure serves as a matching device to overcome the high impedance which occurs at the interface between the transmission system and the phonetic system, and between the phonetic system and the intellect. This mismatch, it is suggested, came about because the transmission system of the vocal tract and the auditory mechanism developed independently of the cortical intellect, necessitating some form of impedance-matching process.

It is pointed out that, not only is the ear not ideally equipped for the processing of speech, but the relationship between the production and reception of sound also falls short of a perfect match (Liberman, Mattingly, and Turvey, 1972, p. 328). Reference is made to Mattingly's proposition that language originates both from the semantic representations generated by the intellect, and also from what he calls the "social releasers" comprised of articulated sounds. Mattingly (1972) suggests that it was the impedance between these two systems which necessitated the evolution of speech grammar. Thus the speech code, by means of several complex conversions, bridges the gap between the acoustic and semantic levels of speech processing.

The distinction between the active versus the passive theories is posited on a direct versus an indirect relationship, respectively, between the acoustic signal and the evoked perception. A direct relationship would not necessitate the existence of a speech grammar. In such an agrammatical system each segment would have to be represented by a discrete acoustic pattern. Unlike the visual system, which processes whole words on a parallel basis, the auditory system receives the acoustic information in serial form. As Liberman has pointed out, the psychoacoustical ability of the auditory system to perceptually process the order of discrete signal units in sound is five to ten times poorer than the rate at which the speech mechanism can generate such units. Following this line of thinking, each unit of a speech signal—treated as a discrete acoustic unit—would require 250 msec. processing time. We know in fact, however, that the rate of normal utterance allows no more than 50 msec. to process each unit (Warren, Obusek, Farmer, and Warren, 1969). Liberman has proposed, therefore, that speech sounds represent a very considerable restructuring of the phonemic "message." This restructuring is necessary in order that the information received at the periphery in as few as eighteen distinctive features (Chomsky and Halle, 1968) can be matched to an intellect comprised of an extremely large number of features identifying differential concepts. Liberman and his collaborators (1972, p. 307) also call our attention to the process of long-term memory. If we are asked to repeat an idea recently expressed by someone we do not duplicate the speech input to our system by repeating exactly what was said; instead we rephrase it. This suggests that long-term memory stores information in a manner very different from short-term memory. Such restructuring, which is the special function of speech grammar, is possible because of the rules of linguistic structure.

Speech perception as a distinctive form.   Let us consider the proposal that speech perception is a distinctive form of perception. We have learned that speech can be produced and perceived at rates well above the limits of the temporal resolving power of the ear. Despite the fact that the ear receives the acoustic signal in serial form, a system of serial processing is clearly not operating in the processing of phonemes. Only parallel processing of the acoustic information at this level could account for such a rapid rate of information flow. This means that the acoustic signal for a phoneme must contain several coexisting bits of information, or constraint criteria, rather than comprise a discrete pattern for that phoneme.

The acoustic cues for successive phonemes are intermixed in the sound stream to such an extent that definable segments of sound do not correspond to segments at the phonemic level. Moreover the same phoneme is most commonly represented in different phonemic environments by sounds that are vastly different. (Liberman, 1967, p. 432)

Yet it appears that rather than proving to be a confusing element, the allophonic variations in the acoustic pattern of phonemes in different phonetic contexts may provide a critical cue to the temporal order of speech events. We noted earlier that speed of processing speech signals is not one of the attributes of the auditory system. Variations in the acoustic patterns of a phoneme in various contexts, however, may well help forecast both context and order of sounds. It has been explained that the articulatory posture involved in the generation of a particular sound is often affected by right to left co-articulation; that is, the anticipation of articulatory movements of subsequent phonemes is evidenced in the acoustic pattern of the phoneme being produced. It is also true that the imprint of the preceding phoneme is to be found in the acoustic pattern of the present phoneme. Liberman, Mattingly and Turvey (1972) illustrates this graphically in Figure 5.3. This schematic depiction of a spectrogram would, if fed to a speech synthesizer, produce the acoustic stimulus we perceive as "bag." The black shapes represent the first and second formants critical to perception of the speech pattern. The influence of each phoneme is clearly seen to extend in a left and right direction to such an extent that the initial and final consonants overlap each other as well as extending into the vowel production area. The influence of the vowel is seen to be effective for the total duration of the word. The extent of the influence of the vowel on the acoustic character of the preceding and succeeding consonants also can be clearly seen. It is evident that a change in the vowel will produce pattern change throughout the sound-pattern of the total word, while the effects of the adjacent consonants are reflected both backward and forward across

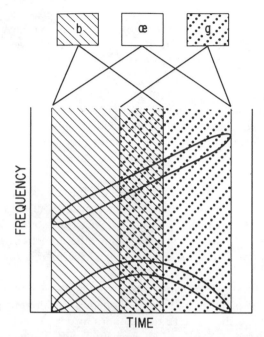

Figure 5.3 Schematic spectrogram showing effects of
coarticulation in the syllable (bæg) (Liberman, Mattingly, and Turvey, 1972, p. 314)

approximately two-thirds of the acoustic pattern. This illustrates the simultaneous (parallel processed) broadcast of several segments of the message.
The effects of certain phonetic combinations are predictable and therefore
highly informative. Thus, not only may the phoneme be identified, but its
place in the temporal order can be determined *precisely because its acoustic structure does vary with phonetic context.*

The form of speech transmission and perception is unique because it
involves the simultaneous transmission of successive segments of the message on the same part of the acoustic signal. This "folding in" of information permits the parallel transmission of several segments of the message.
As we shall see later, the process at the phonetic level probably occurs in
a similar manner at higher levels of language decoding.

Unlocked by a special key. We have already rejected the concept
of the speech code as a system of template matching, since the number of
templates required would need to exceed the total number of syllables in
speech, making it impossible to process speech at speed. The proposal
made by supporters of the active theories is that a mediator serves as the
key to the speech code. To understand this key, we need to recognize the
unique character of speech sounds; namely, that ". . . they are produced by

neuromuscular events that are at some point equivalent to the grammar of the language . . ." (Liberman et al., 1967b, p. 74.) This statement implies a direct relationship between our manner of encoding language instructions for transmission and the corresponding way in which we decode the resultant acoustic signal. The key to the code, according to this line of reasoning, is within the vocal tract: In some way we perceive speech by reference to the manner in which the speech sound was articulated. This theory, termed the *motor theory* of speech perception, proposes that speech sounds represent a code based on the phonemic structure of the language, rather than on an alphabet or a cipher. In a cipher system, a symbol represents each of the units of the original message. In a code, the number of encoded message units is less than the number of original message units. This reduction necessarily results in a restructuring of the information. In their search for the key to restructuring, motor theorists have sought to identify those aspects of the speech signal critical to the perception of particular phonemes. Researchers have been able to isolate the acoustic features critical for, or important to, the perception of most segmental phonemes (Liberman, 1967, p. 434).

It is apparent from the spectrogram for various speech sounds that the features for phonemes are overlapped in the acoustic stream and encoded into syllable-sized units. This process greatly reduces the number of discrete acoustic segments which the ear must process in a given unit of time. It circumvents the problem of the limitations of the auditory system in rapidly processing large amounts of serially ordered information.

As we have seen, there is an almost complete lack of invariance between the acoustic signal and the phoneme perception. Motor theorists have deemed untenable an assumption that the auditory system has evolved a perceptual process based upon the direct matching of the literally thousands of syllables which constitute the language. Research workers hypothesized, therefore, that instead of receiving phonemes directly *embedded in* the sound stream, the phonemes might be *recovered from* the sound stream. This assumption is based upon the concept of encoded information unlocked, or decoded, through a process of reconstructing bits of constraining information transmitted by the source as an acoustic pattern.

Encoding and Decoding. Accepting that speech production and speech perception are two phases of a single communicative process, Liberman et al. (1967, 1967b) trace the perceptual identification of the phoneme back through the acoustic pattern to its origin in the speaker. They have stated:

> We shall assume—indeed we think we must assume—that somewhere in the speaker's central nervous system there exist signals which stand in a one-to-one relation to the phonemes of the lan-

guages. In the act of speaking, these signals, arranged of course in some temporal pattern, flow outward from the central nervous system and eventuate as commands to the articulatory muscles. (1967b, p. 77)

The authors remind us that since we are all both transmitters and perceivers, we all possess this ability to process the segments through the successive stages of restructuring of the information content. We know that the smaller elements of the manner of production of a particular phoneme (those articulatory movements which produce plosion, bilabiality, nasal or oral resonance, and voicing) are constituents of both production and perception. It is proposed, therefore, that it is precisely this information, about *how* and *where* (manner and place of articulation) the acoustic signal was generated, which serves as the key to unlocking the speech code. The invariance, which we determined to be critical to speech perception, while noticeably absent from the acoustic signal, can be found in the manner by which that acoustic signal was produced. Furthermore, the segmentation information necessary for that invariance has been shown to exist, to a considerable degree, at the level of motor commands to articulatory organs.

Thus the motor theory suggests that the speech code is unlocked by running the process backward to recover information about how it was encoded. The encoding process is articulatory. This fact necessitates a key based upon decoding of articulatory gestures which are then restructured up through phonology, syntax, and semantics, to the level of the intellect (Figure 5.4). Liberman and his co-researchers envisage a neurological representation of these features involving "overlapping activity of several neural networks—those that supply control signals to the articulators, and those that process incoming neural patterns from the ear." They further theorize ". . . that information can be correlated by these networks and passed through them in either direction" (1967, p. 454). Thus the act of perceiving someone else's speech is closely related to the process of speaking and listening to one's own speech, with neural activity in the speech mechanism triggered by the incoming auditory data at the level of linguistic features (Cooper, 1972).

Perceived in a special mode.   We have already considered this concept to some degree in our earlier discussion of the neurological theory (Chapter 5, pp. 106–115). We recall our assumption that speech is perceived in a different mode from nonspeech. Lehiste (1972, p. 187) differentiates between these two modes as involving "auditory processing" in the case of nonspeech signals and "phonetic processing" in the case of speech.

As we mentioned earlier (p. 110), one of the differences between the speech mode and the nonspeech mode lies in the fact that the latter involves continuous perception. That is, we track the acoustic variations of pitch, loudness, duration, and timbre. Speech, on the other hand, is perceived by

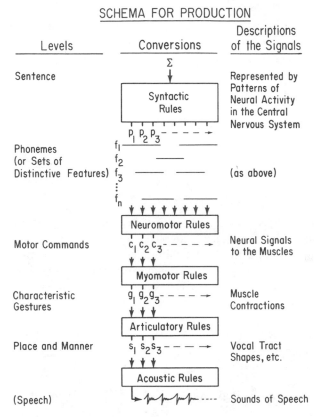

Figure 5.4 Schematic representations of assumed stages in speech production. (Liberman, Cooper, Shankweiler, and Studdert-Kennedy, p. 445. © 1967 by the American Psychological Association. Reprinted with permission.)

means of category boundaries. It has been demonstrated that listeners are able to discriminate variations in phonetic cues better at the phoneme boundary between classes of phones than they can within the class itself. The listener's brain must, therefore, at some point in the process of auditory decoding, differentiate between speech and nonspeech in order to determine how the signal shall be processed. Liberman, Mattingly, and Turvey (1972) reject the suggestion that this decision is based upon the recognition of special acoustic stigmata, or characteristics, associated with speech, as is suggested by the supporters of passive theories. They point out that speech is highly resistant to distortion. Even in highly degraded forms, which would surely destroy any fixed markers, speech has been shown to remain intelligible. In their view the decision is not one made at the early level of acoustic processing but results from phonetic processing. They refer

(Liberman et al., 1972, p. 323) to unpublished data by T. Rand which indicates that the same acoustic stimulus can be processed either as speech or nonspeech. More surprisingly, under certain experimental conditions it has been shown that the brain can process the same acoustic event simultaneously as both a speech and a nonspeech signal. It is suggested, therefore, that at lower levels of auditory processing the signal goes to both speech and nonspeech processors. It is the function of the speech processor to extract phonetic features which identify the signal as speech and gate the neural signals through the appropriate speech processors.

Like the supporters of the neurological theory, Liberman and his colleagues assume the existence of specialized neural units for decoding speech.

> In sum, there is now a great deal of evidence to support the assertion that man has ready access to physiological devices that are specialized for the purpose of decoding the speech signal and recovering the phonetic message. Those devices make it possible for the human being to deal with the speech code easily and without conscious awareness of the process or its complexity. (1972, p. 322)

They disagree with the passive theorists on the question of what the specialized physiological devices process.

Analysis-by-Synthesis Model

Another active theory of speech perception, also based upon the use of information derived from restructuring the articulatory process, is embodied in the model proposed by Stevens and Halle (Halle and Stevens 1959, 1962; Stevens 1960, 1972; Stevens and Halle 1967). They have postulated that

> . . . the perception of speech involves the internal synthesis of patterns according to certain rules, and a matching of these internally generated patterns against the pattern under analysis. We suggested moreover, that, the generative rules utilized in the perception of speech were in large measure identical to those utilized in speech production, and that fundamental to both processes was an abstract representation of the speech event. (Stevens and Halle, 1967, p. 88)

They suggest that the incoming acoustic signal is decoded into an abstract representation of segments and features employing the same set of phonological rules used by the speaker to generate the acoustic signal. This means the listener is processing information, not in terms of the acoustic signal,

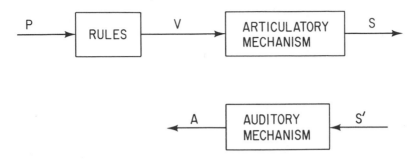

A = AUDITORY PATTERNS
P = ABSTRACT REPRESENTATION
V = INSTRUCTIONS TO ARTICULATORY MECHANISM
S = SOUND OUTPUT

Figure 5.5   Model of the articulatory mechanism and of the speech generating process. (Stevens and Halle, 1967, p. 97)

but in terms of *a knowledge of the rules governing its production*. This immediately surmounts the problem of how variance in the acoustic signal can evoke invariance in the perception. The authors subscribe to the idea of normalization (see our discussion on p. 27). They assume that when a listener encounters a speech pattern different from his own, his auditory perceptual system rapidly computes and defines the difference between the speaker's set of generative rules and his own. He then adds a correction factor to future input, thus normalizing it. In this way the listener rapidly is able to understand dialectical variations not previously encountered. An interesting corollary of this theory is that when a child incorrectly acquires a phonological rule(s), utterances result which are not compatible with the speech patterns of his linguistic environment.

Stevens and Halle use a schematic diagram to represent the skeleton of their model (Figure 5.5).

The authors emphasize that V in the diagram is not representative of the movements of the articulatory structures but is rather "the patterns to be actualized in the form of appropriate sequences of motor commands only after certain motor skills have been acquired." Once the listener knows the rules constraining the speaker (P_____V), he can predict the relationship between the articulatory patterns of the speaker (V) and the resultant acoustic patterns (A) which will be received by his own auditory system." The attraction of the A – V relationship, as described by the authors, is that it means that a direct relationship would exist between the abstract representation and the acoustic pattern. "In this view, therefore, speech would be anchored equally in the motor and in the auditory system of man" (1967, p. 98).

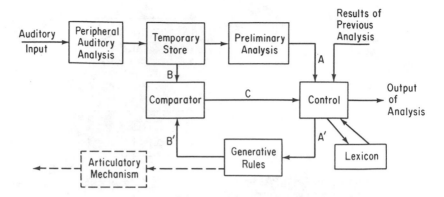

Figure 5.6   Analysis-by-Synthesis Model (Stevens, 1972, p. 50).

The most recent schematic of the analysis-by-synthesis model is shown in Figure 5.6.

In this model, the incoming acoustic speech pattern is subjected to analysis at the lower levels of the auditory system (peripheral auditory analysis). This yields information not only about the frequency and intensity distribution but also about the spectral characteristics of the signal over time. Certain other undefined properties of a non-contextual nature, which can be computed directly from the temporal aspects of the input pattern, are also derived in this preliminary analysis. Together these data represent a partial definition of the pattern of features characteristic of the particular speech sample.

This information is then fed to the master control unit, which also has available to it contextual information computed from adjacent parts of the signal. These are the constraints arising from word sequences and their syntactic and semantic characteristics or features. It has in addition information concerning specific vocabulary constraints plus the output of the comparator. The importance of the contextual constraints has been stressed by Denes:

> . . . the speaker produces acoustic signals whose characteristics are of course a function of the phoneme to be currently transmitted, but which are also greatly affected by a variety of other factors such as the individual articulatory characteristics of the speaker, the phonetic environment of the sound to be produced, linguistic relationships, etc. As a result, the acoustic characteristics of the sound to be produced do not identify a particular phoneme uniquely, and the listener resolves the ambiguities of the acoustic signal by making use of his own knowledge of the various linguistic and contextual constraints mentioned above. (1963, p. 892)

It is the role of the control unit to make an hypothesis concerning the abstract representation characterized by the hypothesized set of units. The hypothesis is stated in terms of generative rules defining certain distinctive attributes which a signal with the hypothesized phonetic representation would necessarily possess. The hypothesis is therefore tested by subjecting it to the phonological rules we use to generate speech—that is, by issuing motor commands to the articulatory mechanism.

In speech perception an inhibitory action blocks this motor pathway. This inhibition results in the production of an hypothetical auditory pattern rather than an articulatory one. The computed auditory pattern is then compared, at lower processing levels, to the input pattern released from temporary storage. If a match is achieved, the information is passed upward, by way of the control unit, for further comparative analysis.

If, on the other hand, a mismatch occurs, the control unit recalculates the probabilities and generates a new hypothesis. Because of the high degree of situational, contextual, and linguistic constraints operating on the speaker, and therefore, to a considerable extent predictable by the listener, the hypotheses formed are generally checked only cursorily.

> . . . we expect that the criteria employed in the matching operation may not always be very stringent. (Stevens and Halle, 1967, p. 100)

Active and Passive Processing: Toward a Compromise

It is important to note that the Stevens and Halle model is not a purely active model. It provides a partial compromise between an internal, generative process and a direct, passive analysis of the signal. The authors stress that they are shifting their emphasis more toward the passive processing of the acoustic signal. In their opinion:

> The relative importance of these two types of processing would presumably depend on the degree to which context must be utilized in decoding the input. . . . (1972, p. 51)

They say in conclusion:

> Our experience with acoustic analysis of speech convinces us that many attributes of the signal for certain features do indeed often remain invariant with context if one permits the inventory of attributes to include certain kinds of time-varying spectral changes as well as simple quasi-static spectral patterns. (Stevens, 1972, p. 51–52)

This apparent compromise between active and passive processing is also to be found in the primarily passive theory proposed by Fant (discussed on p. 104). He states:

My attitude here is basically colored by a faith in the distinctiveness of the speech wave characteristics which we have acquired by being exposed to language in the first place and by reference to our own speech only in the second place. (1967, p. 113)

He also proposes that the motor functions of speech may participate in speech either actively as suggested by Liberman et al. or passively as a secondary activity supporting the direct sensory processing of subphonemic auditory patterns. The author illustrates the interdependence of these two modes of motor participation by the double arrows in his model (Figure 5.7) between the box labeled "auditory patterns" and that labeled "motor patterns." Connections at all stages of processing are shown to exist between the sensory and motor sides of the model.

In writing of the compromise in his model, Fant says:

On the phonemic and subphonemic level I have introduced a block of auditory patterns, CD, in the sensory branch which has its counterpart in a block, GF, of articulations and phonations on the motor branch. A connection between these two blocks indicates the possibility of the participation of motor functions in speech perception either actively, whereby decoding proceeds along the path ABCKFE, or passively in the form of a secondary activity KGHI originating from the block CD while the decoding proceeds directly in the path ABCDE. (1967, p. 113)

He suggests that the ability of a listener to compensate for the loss of large sections of auditory information due to various types of distortion is made possible by "a running synthesis of the most probable continuation of the message, and checking against incoming flow on the sensory side, i.e., along the loop DEFD." This evidences a sensory motor interaction at quite high levels of processing.

Fant concludes:

My guess is that we have means of direct sensory decoding of speech and that the prediction and correction via impulses from the motor system mainly add accuracy in making up for communication deficiencies. In connected speech this prediction is probably indispensable for decoding at a normal speaking rate. (Fant, 1967, p. 115)

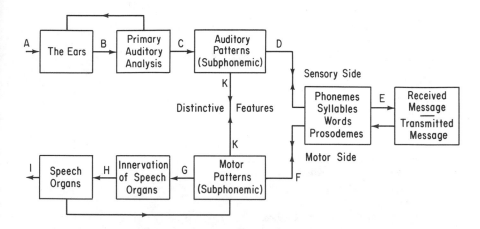

Figure 5.7 Hypothetical model of brain functions in speech perception and production (Fant, 1967, p. 112).

Lieberman, a strong exponent of the motor theory has stated:

The motor theory of speech perception in fact does not really state that listeners even actively compute what the speech signal is by making use of their knowledge of the capabilities of the speech production apparatus. It is quite possible that special auditory receptors exist in man, that are tuned to the humanly possible articulatory maneuvers. (Lieberman, 1972, p. 25)

Denes also writing on the motor theory of speech has said:

. . . just as with many other kinds of human activity, there is probably more than one method by which speech perception can be achieved and normally we may well use several different methods simultaneously. (1967, p. 313)

In conclusion, one can say that it appears unlikely that the perception of speech will be explained by either a purely passive theory or a purely active one. Some combination of the two modes seems to be acknowledged by the exponents of both schools of thought. The solution to the problem of how speech is perceived probably lies not in the answer to the question "Does it involve active or passive processing?", but rather in evidence to indicate the relative contribution of these two modes of function, the variables which govern that contribution, and the levels of processing at which it occurs.

REFERENCES

ABBS, J. H. AND H. M. SUSSMAN, 1971. Neurophysiological feature detectors and speech perception. A discussion of theoretical implications. *J. Speech and Hearing Res.*, **14**, 23–36.

BARLOW, H. B. AND R. M. HILL, 1963. Evidence for a physiological explanation of the waterfall phenomenon and figural after-effects. *Nature*, **200**, 1345–47.

BASTIAN, J., P. EIMAS, AND A. M. LIBERMAN, 1961. Identification and discrimination of a phonemic contrast induced by a silent interval. *J. Acoust. Soc. Amer.*, **33**, 842 (Abstract).

BERRY, M. F., 1969. *Language Disorders of Children.* New York: Appleton-Century-Crofts.

BROADBENT, D. E. AND M. GREGORY, 1964. Accuracy of recognition for speech presented to the right and left ears. *Quart. Exp. Psych.*, **16**, 359–60.

CHANEY, R. B. AND J. C. WEBSTER, 1965. Information in certain multidimensional acoustic signals. Report N 1339, U.S. Navy Electronics Reports. San Diego, California.

CHOMSKY, N. AND M. HALLE, 1968. *The Sound Pattern of English.* New York: Harper & Row.

COOPER, R. S., 1972. "How Is Language Conveyed in Speech?" in *Language by Ear and by Eye*, Part I, eds. J. F. Kavanagh and I. G. Mattingly. Cambridge, Mass.: The M.I.T. Press, pp. 25–45.

CORCORAN, D. W. J., 1971. *Pattern Recognition.* Penguin Science of Behavior Series. Middlesex, England and Baltimore, Maryland: Penguin Books, Ltd.

CORCORAN, D. W. J. AND R. O. ROUSE, 1970. An aspect of perceptual organization involved in the perception of handwritten and printed words. *Quart. J. Exp. Psych.*, **22**, pp. 526–30.

DANILOFF, RAYMOND G., 1973. "Normal Articulation Processes," in *Normal Aspects of Speech, Hearing, and Language*, eds. F. Minifie, T. J. Hixon and F. Williams. Englewood Cliffs: Prentice-Hall, Chapter 5.

DENES, P. B., 1967. "On the Motor Theory of Speech Perception," in *Models for the Perception of Speech and Visual Form*, ed. Wathen-Dunn. Proceedings of a Symposium, November, 1964. Cambridge: The M.I.T. Press, pp. 309–14.

DENES, P. B., 1963. On the statistics of spoken English. *J. Acoust. Soc. Amer.*, **35**, 892–904.

EVANS, E. F. AND I. C. WHITFIELD, 1964. Classification of unit responses in the auditory cortex of the unanesthetized and unrestrained cat. *J. Physiol.*, **171**, 476–93.

FANT, G., 1967. "Auditory Patterns of Speech," in *Models for the Perception of Speech and Visual Form*, ed. Wathen-Dunn. Cambridge: The M.I.T. Press.

FANT, G., 1962. Descriptive Analysis of the acoustic aspects of speech. *Logos*, **5**, 3–17.

FANT, G., 1960. *Acoustic Theory of Speech Production.* 's-Gravenhage: Mouton.

FANT, G., 1956. "On the Predictability of Formant Levels and Spectrum Envelopes from Formant Frequencies," in *For Roman Jakobson*, ed. M. Halle. 's-Gravenhage: Mouton, pp. 109–22.

FRISHKOPF, L. S. AND M. H. GOLDSTEIN, 1963. Responses to acoustic stimuli from single units in the eighth nerve. *J. Acoust. Soc. Amer.*, **35**, 1219–28.

GIBSON, ELEANOR J., 1953. Improvement in perceptual judgments as a function of controlled practice or training. *Psychol. Bull.*, **50**, 401–31.

GOLDSTEIN, M. H., J. L. HALL, AND B. O. BUTTERFIELD, 1968. Single unit activity in the primary auditory cortex of unanesthesized cats. *J. Acoust. Soc. Amer.*, **43**, 444–45.

HALL, J. L. AND M. H. GOLDSTEIN, 1968. Representation of binaural stimuli by single units in primary auditory cortex of unanesthetized cats. *J. Acoust. Soc. Amer.*, **43**, 456–61.

HALLE, M. AND K. N. STEVENS, 1962. Speech Recognition: A Model and a Program for Research. IRE Trans. P.F.I.T., It-8, 155–59

HALLE, M. AND K. N. STEVENS, 1959. "Analysis by Synthesis," in *Proceedings of the Seminar on Speech Compression and Processing*, eds. Wathen-Dunn and L. E. Woods. Vol. 2 AFCRC-TR-5-198, U.S.A.F. Camb. Res. Ctr. Paper D7.

HEMDAL, J. F. AND G. W. HUGHES, 1967. "A Feature Based Computer Recognition Program for the Modelling of Vowel Perception," in *Models for the Perception of Speech and Visual Form*, ed. W. Wathen-Dunn. Cambridge, Mass.: The M.I.T. Press.

HUBEL, D. H. AND T. N. WIESEL, 1963. Shape and arrangement of columns in cat's striate cortex. *J. Physiol.*, **165**, 559–67.

HUGHES, G. W. AND J. F. HEMDAL, 1965. *Speech Analysis.* Lafayette, Indiana: Purdue Research Foundation.

JAKOBSON, R. AND M. HALLE, 1956. *Fundamentals of Language.* Gravenhage: Mouton.

JAKOBSON, R., G. FANT, AND M. HALLE, 1963. Preliminaries to speech analysis. Acoust. Lab., Mass. Inst. Tech., Tech. Rep., 1952, No 13. Cambridge: Republished by the M.I.T. Press.

KIMURA, D., 1973. The asymmetry of the human brain. *Scientific American*, **228**, (12), 70–78.

KIMURA, D., 1967. Functional asymmetry of the brain in dichotic listening. *Cortex*, **3**, pp. 163–178.

KIMURA, D., 1964. Left-right differences in the perception of melodies. *Quart. J. Exp. Psychol.*, **16**, 355–58.

KIMURA, D., 1961. Cerebral dominance and perception of verbal stimuli. *Canadian J. Psychol.*, **15**, 166–71.

KOZHEVNIKOV, J. A. AND L. CHRISTOVICH, 1965. Rech: Articulyatsiai Vospriyatiye (Speech, Articulation and Perception). Moscow-Leningrad: Nanka: Trans. U. S. Dept. of Commerce, Joint Publications Research Service (JPRS). Washington, D.C., No. 30, p. 543.

LADEFOGED, PETER AND D. E. BROADBENT, 1957. Information conveyed by vowels. *J. Acoust. Soc. Amer.*, **29**, 98–104.

LEHISTE, I., 1972. "The Units of Speech Perception," in *Speech and Cortical Functioning*, Chapter 6., ed. John H. Gilbert. New York: Academic Press.

LETTVIN, J. Y., H. R. MATURANA, W. S. MCCULLOCH, AND W. H. PITTS, 1959. What the frog's eye tells the frog's brain. *Proc. Inst. Rad. Engr.*, **47**, 1940–51.

LIBERMAN, A. M., 1957. Some results of research on speech perception. *J. Acoust. Soc. Amer.*, **29**, 117–23.

LIBERMAN, A. M., F. S. COOPER, D. P. SHANKWEILER, AND M. G. STUDDERT-KENNEDY, 1967. Perception of the speech code. *Psychol. Rev.*, **74** (6), 431–61.

LIBERMAN, A. M., 1970. The grammars of speech and language. *Cognitive Psych.*, **1**, 301–23.

LIBERMAN, A. M., F. S. COOPER, K. S. HARRIS, P. F. MACNEILAGE, M. STUDDERT-KENNEDY, 1967b. "Some Observations on a Model for Speech Perception," in *Models for the Perception of Speech and Visual Form*, ed. W. Wathen-Dunn. Cambridge: The M.I.T. Press.

LIBERMAN, A. M., I. G. MATTINGLY, AND M. T. TURVEY, 1972. "Language Codes and Memory Codes," Chapter 13, in *Coding Processes in Human Memory*, eds. A. W. Melton and E. Martin. Washington, D.C.: V. H. Winston.

LIEBERMAN, P., 1972. *Speech Acoustics and Perception*. Studies in Communicative Disorders Series, ed. Harvey Halpern. Indianapolis and New York: Bobbs-Merrill.

LINDGREN, N., 1965. Machine recognition of human language. Part II. Theoretical models of speech perception and language. I.E.E.E. *Spectrum*, **2** (April), 45–59.

MCCAFFREY, A., 1967. *Speech Perception in Infancy*. Unpublished doctoral dissertation. Cornell.

MCKAY, D. M., 1956. "The Epistemological Problems for Automata," in *Automata Studies*, eds. C. E. Shannon and J. McCatthy. Princeton: Princeton University Press.

MATTINGLY, I. G., A. M. LIBERMAN, A. K. SYRDAL, AND T. HALWES, 1971. Discrimination in speech and nonspeech modes. *Cognitive Psychol.* **2**, 131–57.

MATURANA, H. E. AND S. FRENK, 1963. Directional movement and horizontal edge detectors in the pigeon retina. *Science*, **142**, 977–79.

MILLER, G. AND P. NICELY, 1955, 1961. An analysis of perceptual confusions among some English consonants. *J. Acoust. Soc. Amer.*, **27**, 338–52. Also in *Psycholinguistics*, ed. Saporta. New York: Holt, Rinehart, and Winston.

MILLER, G. A. AND W. G. TAYLOR, 1948. The perception of repeated bursts of noise. *J. Acoust. Soc. Amer.*, **20**, 171–82.

MILNER, B., 1962. "Laterality Effects in Audition," in *Interhemispheric Relations and Cerebral Dominance*, ed. V. B. Mountcastle. Baltimore: Johns Hopkins Press, pp. 177–95.

MOFFITT, A. R., 1971. Consonant cue perception by twenty to twenty-four week old infants. *Child Dev.*, **42**, 717–31.

NYE, P. W., 1962. Aural recognition time for multidimensional signals. *Nature*, **196**, 1282–83.

ORR, D. B., H. L. FRIEDMAN, AND J. C. C. WILLIAMS, 1965. Trainability of listening comprehension of speeded discourse. *J. Ed. Psych.*, **56**, 148–56.

STEVENS, K. N., 1972. "Segments Features and Analysis-by-Synthesis," in *Language by Ear and by Eye*, eds. James F. Kavanagh and I. G. Mattingly. Cambridge, Mass.: The M.I.T. Press.

STEVENS, K. N., 1960. Toward a model for speech recognition. *J. Acoust. Soc. Amer.*, **32**, 45–55.

STEVENS, K. N. AND M. HALLE, 1967. "Remarks on Analysis-by-Synthesis and Distinctive Features," in *Models for the Perception of Speech and Visual Form*, ed. Wathen-Dunn. Cambridge, Mass.: The M.I.T. Press.

STUDDERT-KENNEDY, M., A. LIBERMAN, K. HARNS, AND F. COOPER, 1970. Theoretical notes: Motor theory of speech perception: A reply to Lane's critical review, *Psychol Rev.*, **77**, (3), 234–49.

STUDDERT-KENNEDY, M. AND D. SHANKWEILER, 1970. Hemispheric specialization for speech perception. *J. Acoust. Soc. Amer.*, **47**, 574–77.

UTTLEY, A. M., 1959. "Conditional Probability Computing in a Nervous System," in *Mechanisation of Thought Processes*. London: Her Majesty's Stationery Office.

UTTLEY, A. M., 1966. The transmission of information and the effect of local feedback in theoretical and neural networks. *Brain Res.*, **2**: pp. 21–30.

UTTLEY, A. M., 1954. The classification of signals in the nervous system. E.E.G. *Clin. Neurophysiol.*, **6**, 479.

WARREN, R. M., C. J. OBUSEK, R. M. FARMER, AND R. T. WARREN, 1969. Auditory sequence: Confusions of patterns other than speech or music. *Science*, **164**, 586–87.

WHITFIELD, I. C. AND E. F. EVANS, 1965. Responses of auditory cortical neurons to stimuli of changing frequency. *J. Neurophysiol.*, **28**, 655–72.

6

Linguistic Influences on Auditory Processing

We have thus far examined the acoustic signal for its pattern potential. We have seen how the productive mechanism maps patterns on the sound wave and how the auditory system has evolved to break down the complex sound wave and restructure the pattern within the auditory system. We examined certain general perceptual processes in relation to auditory perception, and we considered some specific theories of speech perception. However, we have yet to question the size of the perceptual unit, or to ask how language structure plays a critical role in speech perception, how we pay attention to the auditory signal, or how we store and recall auditory impressions.

It is our goal in this chapter to become aware of these important aspects of auditory processing, to gain some insight into their role in the total process, and to identify some sources of more detailed information. If we can reduce this very large amount of information to its digestible essence, we will have an adequate framework for a consideration of some auditory perceptual disturbances exhibited by children.

Perceptual Units

In the previous chapter, we examined research to indicate that there exist no discrete segments, or units, in the acoustic sound stream (Joos, 1950).

It was shown that, at any given instant in time, information about several phonemes coexists in the sound wave. That is to say, at the acoustic level the signal is not segmented. However, since we perceive it to be segmented, our perceptual system must apparently impose segmentation upon the signal. The next logical question is "What is the size of the segment?" There is probably no single answer to this question, since under different circumstances each linguistic level may be treated as a minimal segment. At times it may be necessary for the auditory system to process the abstract representation of the speech wave phoneme by phoneme in order to reconstruct the message; at other times, the information may be extracted at the level of the syllable, the word, a whole phrase, or perhaps an even larger segment.

To what extent the processing of speech is an inborn human capability remains to be demonstrated. The current research evidence on this topic is reviewed by Morse in the next chapter. If an innate capacity is demonstrable, as the research seems to suggest, then it is necessary to assume that at some level of perception universals exist which cut across all languages. Work by Streeter (1974), Eimas (in press), and Lasky, Syrdal-Lasky, and Klein (1975), which Morse discusses in Chapter 7, provides some exciting initial information on this topic.

However, even if we assume that this intrinsic endowment for perceptual processing of spoken language is a human characteristic, we must still account for the molding of this capacity to accord with specific language structures.

In developing speech perception, we apparently learn to recognize acoustic patterns crucial to our native language. These patterns vary both in length and complexity. When we have acquired the necessary abilities for recognizing a particular pattern, little conscious effort is expended in exercising that ability when the pattern occurs in an anticipated context. In fact, Liberman, Mattingly, and Turvey (1972) have suggested that we probably do little grammatical (phonemic, morphemic, syntactical) processing in normal communication. They suggest that instead we respond in an holistic manner, processing the information summarily as is necessary to verify our predictions of what the pattern *should* be. Consistent interpretation of the analysis-by-synthesis model suggests that processing occurs simultaneously at several linguistic levels, with parallel processing acting within each level.

According to such an hypothesis, when we receive the pattern of an utterance, our mind utilizes knowledge of semantic, syntactic, and phonemic rules to generate an approximation of salient information from the acoustic stimulus. A comparative matching process is carried out, and criteria for acceptance are either fulfilled or unsatisfied. Acceptance occurs within boundaries of tolerance. The rules determining the criteria will differ depending on the level of processing. It is not possible to conceive of an absolute binary choice of acceptance or rejection since we know that the perceptual system shows very broad tolerance within categorical boundaries. Depending upon the strictness of the accept/reject criteria we are able to handle such apparent ambiguities as double entendre, puns, Pig Latin, and other variations in the game of rule bending. Our ability to understand ungrammatical speech also indicates that exact matches are not necessary. Acceptability must, therefore, be conceived of as occurring in degrees and at many levels. The internal approximation (hypothesis) is, therefore, not a random trial but is constrained by the rules of language which determine, in part, what is acceptable. Stevens (1960) suggests that the deductions may be facilitated by an initial scanning of the signal, by coarticulatory information from the spectra of preceding segments, and by familiarity with the input pattern. The influence of context is felt, therefore, at all levels of processing if *the speaker obeys the generative rules of the language.*

The Influence of Grammatical Structure on Segmentation

The influence of grammatical structure on speech perception has been demonstrated by Fodor and Bever (1965) and by Garrett, Bever, and Fodor (1966). These experimenters imposed click sounds on six pairs of tape-recorded sentences at points which constituted natural divisions of the sentences. They then asked the listener to identify exactly where in the sentence the click occurred. The final portion of each pair of sentences was dubbed from a single recording. Thus the beginnings of the sentences were different, but the endings were acoustically identical. Examples of the pairs used in the experiment are:

1(a) In order to catch his train *George drove furiously to the station*

(b) The reporters assigned to *George drove furiously to the station*

5(a) As a direct result of their new invention's *influence the company was given an award*

(b) The retiring chairman whose methods still greatly *influence the company was given an award*

Note that the italicized sections are acoustically identical, check marks indicate the positions of the clicks.

The perception of the click seemed to shift toward the natural sentence break, which was different for each sentence. That is to say, of the two click positions, the one perceived by the listener was biased by the particular structure of the sentence involved. This indicates that at least under these test conditions, grammatical composition alone causes us to segment the signal differently for different linguistic structures.

It would appear that a direct relationship also exists between the pattern of the sentence structure and its intonation pattern. In 1967, Lieberman proposed a theory which links together intonational (suprasegmental) processing and syntactic analysis. He maintains that the basic function of intonation is to establish constraints to permit identification of segmental, linguistic units.

. . . the primary linguistic function of intonation is to furnish cues that allow the listener to segment speech into blocks for syntactic processing. A syntactic recognition routine is the prerequisite for the semantic interpretation of the speech signal. The listener must derive an underlying phrase marker (see Katz and Postal, 1964) starting with the acoustic signal and his knowledge of the grammar of the language. The underlying phrase marker is the input to the semantic component of the grammar. Intonation furnishes acoustic cues that tell the listener when he has a block of speech which constitutes a satisfactory input to his syntactic recognition routines. Intonation can furnish different meanings to utterances that have the same words by grouping the words into different blocks which direct the listener's recognition routines toward one underlying phrase marker rather than another. (1967, p. 315)

Martin (1972) has likewise proposed the existence of a definite relationship between grammatical composition and acoustic pattern. In his opinion, the rhythmic structure of speech constitutes the temporal patterning of the acoustic signal. Rhythm arises from the natural movement sequences involved in speech production, influenced among other things by the constraints of timing. That is to say, the motor movements are constrained by the natural limits of the ability of the articulators to function in time (rate of movement, change of movement, rate of change, coordination of movements). Martin maintains that the influence of these constraints is reflected in the organization of sounds.

In the case of speech, these constraints influence the segmental details of the acoustic signal, but they should be expected also to affect at least some aspects of the morphology and syntax of any language at

> the level of syllable strings since they play a role in determining what can be naturally and easily spoken. The constraint on speech sounds, or on any other real-time sequence of behavioral elements, that is directly implied by the concept of rhythm is *relative timing*, which means that the locus of each (sound) element along the time dimension is determined relative to the locus of *all* other elements in the sequence, adjacent *and* non-adjacent. That is to say that sequences of sounds, speech or otherwise, that are rhythmic will possess hierarchical organization, that is, a coherent internal structure, at the sound level. (1972, p. 488)

The author suggests, therefore, that relative timing during speech production determines temporal location within the utterance and, therefore, the duration of each sound element is related to each other locus in the resulting pattern.

He explains that the constraints placed by rhythm on the production of speech mean that the sound inputs during perception are *temporally* patterned. He goes on to state:

> Furthermore since rhythmically patterned sounds have a time trajectory that can be tracked without continuous monitoring, perception of initial elements in a pattern allows later elements to be anticipated in real time. (1972, p. 488)

Martin provides a possible explanation of how the higher units of perception may be linked to the acoustic stimulus. The constraining effects of rhythmical patterning are imposed upon the speaker by the dynamics of the physiology of speech production. The rhythmical patterning of speech, in which the syllables are timed relative to the total pattern rather than simply presented in serial order, suggests a preset timing of speech under central control. The hierarchical theory proposed by Martin hypothesizes that:

> . . . since the accented elements dominate the temporal organization they must in some sense be planned first. Intervening, lower-level syllables then are planned subsequently in hierarchical fashion, by (metaphorically) reading the rhythm tree, level by level, from the top down. . . .

> In this view one might think of accented syllables as the main targets in the organization of the articulatory program. (1972, p. 499)

Since there is only a certain amount of time available for the utterance of a speech sound in rapid speech, Martin suggests that "the accented syllables are articulated at the expense of lower-order syllables. . . ." As the number of syllables between stressed syllables increases, the phonetic detail of the unstressed syllables becomes progressively more blurred and less precise. However, since the temporal patterning loads the stressed syllables with information, redundancy is built into the system, decreasing the need for detailed analysis of the unstressed syllables.

According to Martin's theory, the perceptual system is cued by the rhythmic key to identify, through preliminary analysis, those sections of the utterance containing the greatest information value. Information from the lower levels, interpreted in a secondary analysis, is subject to the constraints obtained from the primary analysis. This is possible, the author suggests, because the rhythm pattern information (stage one) can be processed independently from segmental information (stage two).

Martin points out that this two-stage analysis is not dissimilar from the analysis-by-synthesis model of Halle and Stevens. In processing speech sounds the coarticulatory information permits prediction concerning how the dynamics of the speaker's utterance will probably develop.

> Hence it is not simply, or not only that discrete arrival times of accented syllables are induced from earlier timing relationships but also that the total array of time-varying cues in the continuous flow of speech will project ahead the general outline of the remaining prosodic contour. These cues telegraph not only tempo changes but more generally the whole thrust of the pattern of sounds yet to come. (1972, p. 503)

The computing of expectancies in this manner allows for the advanced gating of the auditory system to facilitate rapid processing of the signal within the expectancy constraints generated from the preliminary temporal analysis. Providing a quick scan confirms the prediction, no deeper analysis is necessary, and the extent to which the actual signal needs to be processed is markedly reduced.

The Levels of Segmentation

Lehiste (1972) has provided an extensive review of the experiments directed at establishing at which levels perceptual units may indeed be established. She also examines the evidence pertaining to the role of stress in the perception of units at the sentence level.

In considering segmentation at the subphonemic level, Lehiste discusses the ability of the auditory system to perceive both allophones (variations within the phonemic category) and the distinctive features of phonemes. While rejecting the claim that the context-sensitive allophone (Wickelgren, 1969) constitutes the basic unit of perception, she acknowledges that both duration and manner of articulation can be perceived by the listener. She agrees that the *minimal* elements of speech perception must be located at the subphonemic level (p. 197). This minimal level does not, however, preclude processing of larger sized segments. Reviewing the evidence of the system's ability to process perceptual units at the phoneme level leads her to conclude:

> It seems that a level of perception at which phoneme-like units are responded to should be recognized; it remains to relate it to the other levels of perception for which evidence has likewise been provided by studies of speech perception. (1972, p. 199)

The evidence concerning the processing of syntactic units has been provided by the work of Fodor and Bever (1965) and Garrett, Bever, and Fodor (1966), which we discussed earlier. Similar evidence has been provided in studies by Bever, Lackner, and Stoltz (1969); Bever, Lackner, and Kirk (1969); Bever, Kirk, and Lackner (1969). The results of each of these studies leads Lehiste to the conclusion that

> . . . listeners use grammar actively to impose syntactic structure on the speech stimulus as they hear it. Listeners respond in terms of the underlying structure of the sentence rather than its surface structure. (1972, p. 211)

Lehiste goes on, however, to discuss further studies (Abrams and Bever, 1969; Holmes and Forster, 1970; Bond, 1971) which do not support each of these claims. The conflicting evidence, she concludes, indicates the existence of two levels of processing: a primary process and a linguistic process. *Primary processing* is equivalent to auditory and phonetic analysis; i.e., listening in the speech mode. *Linguistic processing* (phonological and syntactic analysis) is dependent upon preliminary auditory analysis, though Lehiste concedes that phonetic and linguistic processing may at times take place concurrently. The size of the segments processed, it is suggested, may differ at the various levels with extensive interaction among them. However, syntactic analysis is seen as being dependent upon prior phonetic analysis (p. 228).

Neisser agrees that the auditory system is able to segment the signal into units of differing size. He says:

> Auditory synthesis, like its visual counterpart, can apparently produce units of various sizes. The listener can ask himself, "What sounds are uttered?" or "What words were spoken?" or "What was meant?" and proceed to synthesize accordingly. In each case he must have a set of rules: phonetic, phonemic, syntactic, semantic, or what you will. (1967, p. 194)

The role which a knowledge of these rules plays in speech perception cannot be too strongly stressed. Garvin (1971), in an article comparing human and computer languages, emphasizes that the two differ significantly in the matter of predictability. While computer language is highly regular, programmed on a purely binary system, human language has only partial regularity (the exceptions often seem to outnumber the grammatical rule!) and is highly complex in organization. He explains:

> The human language system can be characterized as a multiple hierarchy, which means that the structure of language is organized in terms of more than one structural level and more than one organizing principle. (1971, p. 283)

Pierce (1969) also stresses that speech perception is heavily dependent upon familiarity with the rules of language processing. He points out that computer recognition of speech is made particularly difficult because *the human listener depends very little upon the information in the acoustic signal and very heavily upon his knowledge of the linguistic rules constraining the speaker.* As Lehiste states:

> Acoustic cues alone do not determine the boundaries of the perceptual units. (1972, p. 211)

The Ordering of Phonemic Units

The perceptual, or cognitive, units we have been discussing are reconstructed from the spatiotemporal information contained within the acoustic signal. This is true regardless of whether we are dealing with the morpheme or the sentence. The true meaning lies not simply in the identity of the component segments but, more importantly, in their relationship to each other, that is, in their temporal order.

It is obvious that the phonemes /t/s/p/i/ arranged as p i t s constrain one's thought in a different manner from the same phonemes arranged as s p i t. The sequential order of phonemes is therefore critical in our restructuring task.

Neisser (1967) points out that the 20 to 30 msec. duration of the sound segments is very close to the minimum time within which we can perceive temporal order (Hirsh, 1959), evidence of which, he says, is seen in the not infrequent reversal errors made by young children. My own children, for example, generated such words as:

/baikɪsl/	for bicycle
/sɪrəgɛt/	for cigarette
/frɪrɪdʒərɛitə/	for refrigerator

Neisser also refers to a study by Broadbent and Ladefoged (1959) in which they compared the ability of subjects to discriminate the order of certain synthesized consonants. It was shown that a greater number of errors were evidenced by subjects unfamiliar with the particular sounds used than by subjects who were experienced. The experienced subjects stated that they listened for a quality difference between the sound pairs rather than an order difference. Neisser concludes from this and other studies that:

> It does not seem that the listener isolates /s/ and /t/ before examining their order. Instead, he gradually acquires a capacity to distinguish /st/ from /ts/. (1967, p. 184)

It would seem, again, that the perceptual system perceives certain consonant clusters to comprise holistic units rather than simply treating the cluster as a series of phoneme segments.

Lehiste (1972, p. 200) has reviewed studies by Bond (1971) and Day (1970a, 1970b) each of which was concerned with the apparent unitary nature of the perception of consonant clusters. In the Bond study, subjects discriminated between fifteen word pairs differing only in the phoneme order of the consonant pairs /ps–sp/, /ts–st/, and /ks–sk/ (e.g., task–tax, lisp–lips, coast–coats) presented under poor signal-to-noise-ratio conditions. Correctness and reaction times were studied. It was found that the subject took longer to provide a response which proved incorrect than to give one which subsequently proved to be correct. This indicated that when the pattern was familiar, less time was consumed than was demanded by unfamiliar combinations processed phoneme by phoneme. When set to perceive a whole pattern, an unfamiliar combination was generally incorrectly processed. Reversal of the consonants was the most common error, further suggesting that the stimulus was perceived as a whole unit rather than phoneme by phoneme. This observation is in agreement with the opinion of Neisser quoted above.

In the Day studies, subjects were asked to report on two stimuli presented simultaneously to opposite ears. The stimuli differed in the initial consonant (e.g.,/bæŋkət/ and /læŋkət/) and were presented with differing onset times ranging from 25 to 100 msecs. The subjects always reported hearing the word blæŋkət regardless of which of the two actual stimuli reached an ear first. This demonstrates how strongly linguistic constraints influence auditory processing according to the rules of the listener's language system. In the second study by Day, it was shown that providing the reversal of sound order permits an acceptable word to be identified (e.g., past/pats), temporal order judgments can be made equally well for both patterns.

Once again, then, the evidence points to pattern rather than discrete phoneme identification. These patterns are distinctive by virtue of the linguistic rules which govern their generation by the speaker, rules which operate from the phoneme level to the level of syntax.

Ordering of Word Units

In the previous chapter (p. 118) we considered Mattingly's proposal that speech grammar serves as the bridge between the acoustic and semantic levels of speech processing. Grammatical rules at the level of the sentence (syntactical rules) operate in a manner similar to those limiting phonemic sequencing. Syntactic rules govern the possible relationships between the words in the sentence and the relationship of each word to the total sentence pattern. As we observed at the phonemic level, these structural rules go beyond determining the simple order of the words. They reflect interrelationships in the same sense as the phoneme which has acoustic influence well beyond its own particular time section of the event. We see this intricate patterning repeatedly at each of the levels in the structure of the message signal. This process can be compared with those intriguing anatomical atlases, in which transparency is superimposed upon transparency in a layered order to reveal increasingly complex relationships between the deep structures. Each level is related in an ordered manner to the one above and the one below; each is also related to the whole and, therefore, indirectly to all other levels.

The ability to comprehend sentences as holistic units rather than serial-ordered word units is dependent upon the listener's unconscious knowledge of the rules by which the speaker generates them. For each language, basic relationships exist between the component parts of different types of sentence functions. The syntax of the grammar reveals what particular relationships cause to be predicted; i.e., how they function. The surface structure of sentences can be considered to have a graded order. The sentence [*The boat sank.*] comprises two components, a *subject* and a *predicate*. In linguistics, symbols are used to represent these grammatical labels. The annotations are descriptive of the nature of the component which they identify. For example:

Complete sentence	S
Subject	NP (noun phrase)
Predicate	VP (verb phrase)
Object	NP (noun phrase)
Article	DET. (determiner)
Tense	T.

Thus our sentence can be diagrammed:

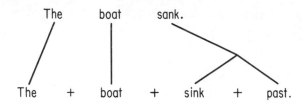

```
   The        boat      sank.
    /          |          >
   /           |         / 
  The    +    boat   +   sink   +   past.
```

In the case of a slightly more complex sentence, [*The boat leaked water.*], the components would be: Subject + Predicate + Object. This is diagrammed as follows:

The boat leaked water

```
                           S
                ┌──────────┴──────────┐
               NP                     VP
             ┌──┴──┐              ┌────┴────┐
           DET      N            VP          N
            |       |          ┌──┴──┐       |
            |       |          V     T       |
           The     boat      leak   past   water
```

These diagrams of the surface structure of very simple sentences reveal the role of the words in the sentences and the relationship of the words to each other. It is the surface structure which identifies the meaningful units of speech by indicating which groupings are intended.

The syntax consists of:

1. a set of base structural rules
2. rules for transforming a basic sentence from one form into another form; e.g., negative, interrogative, etc.
3. morphological rules.

As I have said, the structural rules refer to the *relationship of the roles of the sentence components*.

Sentence (S)	Noun phrase (NP) + Verb phrase (VP)
Noun phrase (NP)	Determiner (article, adjective, etc.) + Noun
Verb phrase (VP)	Verb (V) + Noun phrase (NP)

Each sentence is constructed from a particular *kernel*, or nucleus, by applying rules such as we have listed. As Menyuk (1971, p. 17) explains, each component of the sentence rules (N, V, NP, VP, etc.) has certain phonological, syntactic, and semantic characteristics. Furthermore, on the basis of these properties, *constraints are imposed which limit the sequential relationship of the components in the string*. These constraints are responsible for the grammatical pattern. Menyuk goes on to explain that transformational rules have been employed to convey information about the arrangement of the base structure components, since these must also fit within the constraint pattern. She points out that "A bundle of syntactic and semantic properties and phonological features marks *each item* in the string" (p. 18). As an example, she provides a partial description of the underlying base structure of the sentence [*The boy hits the ball*] (Figure 6.1). In this example, the properties embodied by the item *boy* are: noun/singular/animate/human/male, etc. The speech sound features for /bɔi/ are also part of this item's character. In describing transformational rules, Menyuk states:

> Transformational rules are rules for operations on underlying strings to derive various sentence types. By addition, deletion, permutation, and substitution within or among kernels, various sentence types are derived. Morphological rules are rules for the application of grammatical markers such as person, number, tense, etc. (1970, p. 18)

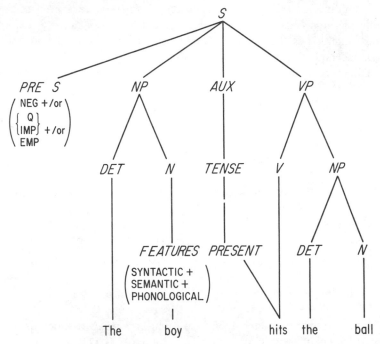

Figure 6.1   A partial description of the underlying base structure of the sentence "The boy hits the ball." (Menyuk, 1971, p. 18)

She provides the following description of how the application of transformational rules may effect changes in the underlying kernel.

1.	Underlying string:	The boy present hit the ball.
2.	Negative placement:	The boy present not hit the ball.
3.	Do Support:	The boy present do not hit the ball.
4.	Q. Aux. Inversion:	Present do not the boy hit the ball.
5.	Affix hopping:	Do + present not the boy hit the ball.
6.	Contraction:	Do + present + nt the boy hit the ball.
7.	Morphological rules:	Do + present + nt : Doesn't
8.	Surface string:	Doesn't the boy hit the ball?

Over and above the constraints inherent in the syntactical structure, the thought transmitted has been subjected to semantic constraints. Thus, the words may satisfy syntactical criteria as in the sentence "Colorless green ideas sleep furiously," yet make semantic nonsense.

The point to be made is that language is highly structured in an hierarchical, or layered, form. The development of a sentence at the phonetic level, the highest or most surface level of the hierarchy, is achieved by a series of structural transformations which involve the "folding in" of information. We have already considered this folding in process in the transforming of information from the acoustic level to the phonemic level (p. 120). The linguistic restructuring thus involves:

1. the transformation of the information in the acoustic event received by the cochlea into perceived phonemic information
2. the use of phonological rules or constraints to restructure the *surface structure* of the sentence
3. the use of syntactical constraint information to reconstruct the deep structure which permits meaning to be attributed.

Liberman uses a tree diagram to indicate this restructuring process at the syntactic level (Figure 6.2).

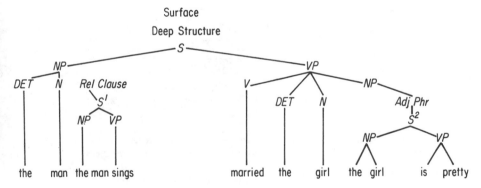

Figure 6.2 A tree diagram showing how the segments at one level (deep structure) are organized into larger units. (Liberman, 1970)

In order to simplify the structure, the process involves the rearrangement, grouping, and elimination of components. If the listener knows the rules of the process, his language processing system will be able recover the deep structure and therefore the meaning from the surface structure.

Liberman has illustrated this process thus:

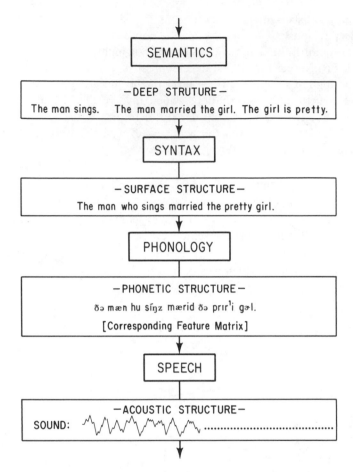

Figure 6.3   An illustration of the assumption that the sounds of
speech are a separate level of language, connected
to the phonetic structure by a grammatical recoding
similar to syntax and phonology. (Liberman, 1970,
p. 302)

We can summarize this very brief glimpse at the role of linguistic
structure on the perception of speech by placing it within the framework
of pattern processing and recognition. This framework has been the unify-
ing concept in our examination of the auditory processing of the speech
signal. To recapitulate: We have seen repeatedly that the acoustic event
itself does not carry meaning. Its role in speech communication is to provide, in patterned form, the information necessary to permit the listener
to recreate the intended message. The acoustic pattern was brought into

existence as the result of information which was subjected to several transformations during the production processes of speech. It is logical to expect
that in some way those transformational processes will have to be run
backward if the listener is to be able to recover, or reconstruct, a satisfactory equivalent of the speaker's thought. In the final event the actual
meaning, the *value* of interpretation the listener assigns to the reconstructed
message, will depend upon the degree of compatibility between the experiences of the speaker and listener. However, if we leave value aside, competence in restructuring the total pattern of the message depends upon the
listener's knowledge of the rules which govern the preparation of that
message.

Our earlier examinations have revealed the nature of the recovery
of the phonetic structure from the acoustic pattern. In this chapter, our
brief look at the linguistic structure of the message has indicated that it
too is highly patterned, and that the pattern is assembled by a series of
restructuring transformations which follow well defined rules. Liberman
(1970) and Liberman, Mattingly, and Turvey (1972) have explained how
the conversion process, studied experimentally at the phonemic level, may
apply to higher level linguistic processing. The structural rules governing
the decoding of speech and the rules for deriving the phonetic units from
the sound are integral components of the same system, namely the grammar of speech. Or, as Liberman states:

> . . . the interconversion of phonetic segment and sound is a gram-
> matical recoding, similar in complexity and form to syntax and
> phonology . . . speech is truly an integral part of language, not
> merely a convenient vehicle for transmitting it. (1970, p. 304)

Memory Process

Throughout our discussion of speech perception, we have emphasized
that speech is a spatiotemporal phenomenon: The speech wave is generated by the modification of vibrated air passing through the vocal articulatory tract. Changing articulatory spatial relationships pattern the acoustic event and produce a speech wave which occupies space and behaves in
a manner varying over time. The spatial distribution of energy along
the basilar membrane of the organ of Corti represents spatial information; this distribution changes over time, thus reproducing the spatiotemporal pattern. We also demonstrated that the resultant neural representations are based upon spatial patterning within the nervous system
and on the temporal order of the triggered impulses. Finally, we saw the

role that the temporal relationship of the component parts plays in determining how segments are assembled.

Speech is a continuous phenomenon. Not even the smallest segment
is available in its entirety at one instant in time. It is obvious, therefore,
that before it can perform analysis with any degree of sophistication, the
auditory system must wait until sufficient information has entered to permit
the processing of the smallest sized unit. Even a single speech sound takes
time to produce. Its perceptual identity will depend upon information
about the manner of its production or, in the case of passive analysis, information about the acoustic pattern resulting from its manner of production. Larger segments are even more dependent upon the passage of real
time to allow for the buildup of patterned information.

It is apparent, therefore, that the system is capable of retaining the
peripherally processed information, as it comes from the cochlea, for a
period long enough to identify at least the smallest units of the pattern.
Norman (1972) has suggested that there may initially exist a very low
level of internal recording equipped to hold, for brief periods, information
about the acoustic waveform internalized as an equivalent sensory waveform. This sensory information imprinting involves fast fading, continuous,
serial-ordered information referred to by Neisser (1967) as Iconic memory.
At this stage the information is held just long enough for some small
degree of restructuring to take place, probably involving the extraction of
feature information of a precategorical nature which Crowder (1972, p.
261) identifies as pitch, voice quality, location, and loudness.

This information, it is suggested (Norman, 1972), then passes into
what has been labeled P.A.S., *precategorical acoustic storage* (Crowder
and Morton, 1969), or *echoic* memory (Neisser, 1967). The information
is held here for an estimated one second (Guttman and Julesz, 1963) to
6–8 seconds (Cole, Coltheart and Allard 1974), and ten seconds (Eriksen
and Johnson, 1964). However, most studies (Crowder and Morton 1969;
Bjork and Healy, 1970; and Estes 1970) place the retention span of echoic
memory at 2 to 3 seconds. This period is necessary to permit the buildup
of information necessary for the extraction of linguistic features and the
essential suprasegmental information identifying stress and intonation
patterns.

Short-term memory. A further encoding, or restructuring, takes
place converting the information from continuous echoic form to segmented linguistic form, either in terms of articulatory or acoustic distinctiveness. Estes (1972, p. 178) has suggested that the information at this
level is held in an hierarchical form using a parallel system, with each
indicator responsible for a particular characteristic of the stimulus code.
This concept of mapping of distinctive-feature information onto specialized

neurophysiological structures of the auditory system is familiar from our earlier discussion (Chapter 5, p. 104). We also discussed the concept of parallel processing of sensory information.

The information thus processed is held in storage in a *short-term memory* system in an organized segmented form. It is important to note that the restructuring has completely changed the form in which the information exists. Neisser stresses the difference between echoic and short-term memory thus:

> One is passive, the other active; one is continuous, the other segmented; one is composed of sounds, the other of speech; one seems to decay rapidly, the other can be renewed indefinitely through rehearsal. (1967, p. 226)

Short-term memory is therefore the first level at which we can exert conscious control over memory content.

Long-term memory.   Berry (1969, 107) has stated, "Memory is the retention of patterns of perception." This applies to each level of memory. It is the retention time and the character of the restructured information which differs. If larger patterns of information, and the values assigned to them, are to be available to the perceiver, it is necessary that the processing system provide retention of a much longer period. This task, fulfilled by long-term memory, makes possible the retention of information about the complex phonetic and syntactic relationships within the speech signal.

The fact that the information in long-term memory is segmented has to do with the number of units it can hold at a time: Long-term memory is capable of holding between 5 and 9 units of a stimulus. Miller (1956), in a paper entitled "The Magical Number Seven Plus or Minus Two," indicated that this limitation operates even when we are remembering strings of numbers, words, or, in fact, any kind of units. It appears that we surmount this limitation by grouping the components into what Miller calls "chunks," or cognitive units created by the listener. This must occur since we obviously are capable of remembering sentences of more than nine words in length. "Chunking" is a process of segmenting or clustering units into groups. When we use this method the limitations of a seven-unit memory span are circumvented, for we are now able to store seven chunks. It is of interest to note that this 7 ± 2 finding holds true for various types of stimuli within a sensory modality and also across sensory modalities. It also seems to correlate with the idea of the segmentation of speech into variable-sized linguistic units (p. 141); indeed we could delineate the "chunks" by the stress patterns of speech.

We have tried to avoid the use of the terms "store" and "storage."

Current thinking rejects the concept of memory as a neurophysiological store in which images or traces, termed *engrams* (Lashley, 1950, pp. 477–79), are accumulated. Instead we have projected an auditory perceptual system which is constantly being adapted to an anticipated signal by a continuous tuning process. More compatible with this approach is a concept of memory involving the patterns of potential interaction between cells. These gating, or tuning, mechanisms (p. 71) allow for the variable retention times of short- and long-term memory by varying the transmission potentials of the neural units involved (Eccles, 1953). The establishment of long-term memories results then from structural changes of nerve cells, changes that alter their potential for interaction with other cell units (Fair, 1965). In this way, a particular spatiotemporal pattern is more or less permanently retained by a pattern of neuron activity. This activity has a resistance to fading equivalent to the degree of learning which has occurred (cf. with echoic memory). Commonly used patterns would be immediately identifiable at an automatic preconscious level analogous to automatic sequential behavior. Less routinely evoked patterns would have a somewhat lesser degree of permanence reflected in the greater amount of attention necessary to process them.

Selective Attention

The more permanent the internal pattern representation is, the less dependency on the external information contained in the acoustic event. If asked to listen to a list containing familiar items, e.g., days of the week, the multiplication tables, or a well-known verse, detailed memory permits rapid identification of the pattern of the list—and prediction of future items. Therefore, the acoustic signal becomes highly redundant. The same is true to a lesser degree if a fairly long list contains only items that fall into a certain category, i.e., flowers, presidents of the U. S., staple food items, etc.

For this list we must focus attention somewhat more sharply on the external stimulus, which no longer has the same degree of redundancy. This is necessary because the same degree of patterning cannot be imposed on the overall list as we did when listening to the closed-set lists of the days of the week or the multiplication tables. The probabilities of occurrence of any potential item are greatly reduced by the size of the category. La Berg (1962) suggests that the procedures for processing information into echoic and long-term memory involve high levels of attention when new tasks are undertaken. However, he believes that sufficient practice results in the relegation of the task to a relatively automatic procedure releasing the system for high level processing of other information patterns. He stipulates that the same process operates in developing expectancies which lead to the identification of "the proper perceptual settings" (p. 246) and the matching of external information to the internal model. This process is probably also modified through experience until it becomes relatively automatic.

> Thus the change in amount of attention needed to process information flowing outward from memory to perceptual levels over the course of learning is analogous to the change in attention requirements during learning to process information flowing inward from receptors to memory systems. (La Berg, 1972, p. 247)

This relegation of processing to preconscious stages releases the neural system from concern with how to deal with incoming linguistic information, whether it be in spoken or written form. It permits concentration on the higher processes of determining semantic values and developing more sophisticated predictions. The fact that strategies are involved in lower level (i.e., linguistic) processing may be hard for the average person to recognize. For the child with speech perception problems (and often with resulting speech and language production problems), or for the child experiencing difficulties in the acquisition of reading skills, however, such tasks are frequently not relegated to preconscious processing. If we are to help individuals with these difficulties, we will have to understand the process by which strategies are learned and automated.

We have held that speech perception constantly attunes the incoming information according to probability computations to increase internal redundancy, decrease dependency upon the external stimulus information, and thus reduce to a minimum the actual perceptual processing time necessary for reconstructing the message. La Berg, strongly supporting this treatment of speech perception, writes:

> What we are becoming slowly aware of in our attention experiments is that there must be important information flows from higher processes outward to the perceptual levels that allow the organism to tune his ear or eye to the sort of stimulus he expects, and thereby significantly cut down his processing time from what it would be were he to receive information at a constant state of perceptual preparation, that is, in a linear, take-it-as-it-comes manner. (1972, p. 247)

The problem of focusing attention is that, in order to attend to the desired message signal, it first must be identified. We choose from among the multitude of simultaneous acoustic events that contribute to the overall complexity of the acoustic pattern impinging upon the eardrum. Broadbent (1958, 1962) has suggested that a filtering system operates to select those channels of information which shall permit an upward flow of processed data. A tuning process selects the channel dealing with information patterns of anticipated relevance and blocks off those channels whose patterns are not currently appropriate.

A modification of Broadbent's theory advanced by Treisman (1960) is more compatible with our approach to perception. In the Treisman model, a relative tuning mechanism rather than a simple open-shut device is proposed. The degree of attenuation imposed upon flow paths considered less than maximally relevant would be dependent upon expectations based upon constraint information from past experience or from early processing stages. Such a model allows for an hierarchy of relative values, with the greatest concentration of attention on the most relevant information. Other patterns are processed simultaneously but with less detail. The gradients in the system are determined by compiled probabilities and are achieved by neural gating.

Neisser suggests quite a different explanation (1967, p. 212–18). He proposes that auditory attention operates through the process of analysis-by-synthesis. He explains that our ability to pay attention to one conversation in the presence of others, or indeed to listen to the melody of a piece of music to the exclusion of the accompaniment, results from the fact that only one external pattern will closely approximate the internally generated model against which it is matched. He states "I suggest *that this constructive process is itself the mechanism of auditory perception.*" The ability to follow a particular speech message depends upon synthesizing, according to linguistic constraints, a pattern of linguistic segments which satisfactorily matches the incoming pattern. This then becomes *the figure* which is processed upward through the higher levels. The remaining pattern information, *the ground*, just simply does not get processed beyond the preattentive stage. This ability to process stimuli at preset levels, determined either by analysis-by-synthesis, or by variable tuning, has been demonstrated (La Berg, 1971a, b). La Berg has stated:

> . . . the subject may tune his perceptual analyzers according to the sort of stimulus he expects to receive, and, in addition, he can adjust himself to process that stimulus at a chosen level. (1972, p. 245)

La Berg has suggested that the procedures for processing information up from echoic into short- and long-term memory require high levels of attention when new tasks are involved but diminishing attention levels as the task becomes more familiar. This concept of applying different strategies and different degrees of attention to new learning tasks versus well-learned behavior has implications for those concerned with developing procedures for training or retraining children with auditory perceptual problems affecting the acquisition of speech, language, or reading skills.

In summation, the influence of linguistic structure on speech perception cannot be overestimated, since it is operative at all levels of processing from the phoneme to the sentence. It is the language rules, operating on speaker and listener alike, which govern how we recover the coded instructions from the acoustic signal. Furthermore, because linguistic rules govern the continuous generation of predictions about how the acoustic signal will evolve, it is possible to tune the auditory system to facilitate the reception and processing of the anticipated pattern. Such facilitation enormously reduces the dependency of the perceptual system on the acoustic signal itself by generating high levels of internal redundancy. Without this, we would not be able to process speech at such a rapid rate.

We have seen that such a dynamic system as exists for speech perception is capable of shifting its level of perception from treatment of the details of subphonemic units to the processing of large syntactic units. We recognized that two levels of processing probably coexist: primary processing of auditory information, which is the act of listening in a speech mode (Chapter 5, p. 124), and secondary processing, which is linguistic in nature. This second stage involves the segmentation of the continuous stream of acoustic information into units of varying size at different levels of processing (phonetic, phonemic, syntactic, semantic). At each level, a set of rules appropriate only to that level is used to decode the information and restructure the message. The particular rules we use necessarily influence what we perceive.

We saw that the identification of the particular grammatical composition of the message is probably cued, at the primary processing level, by the temporal pattern of the acoustic signal (the suprasegmental information).

Berry states this clearly:

The perception of syntactic and semantic cues hinges primarily upon the earlier perception of intonational patterns, for they furnish natural guides to these linguistic features. (1969, p. 49)

Like Martin (1972) whose concepts we discussed in some detail, Berry believes that the various segmental cues must be mutually consistent with these more basic intonational patterns emerging from archetypal breath-groups. Thus the primary and secondary levels of auditory processing of speech are complementary.

We traced the influence of these language rules up through the syntactic level involving the perception of whole units rather than serially ordered words. In so doing, we learned that our perceptual system imposes sentence structure on the acoustic information as it is received. Furthermore, we are influenced not only by the words of the sentence but by a knowledge of the relationships among them and the function of those particular relationships. This knowledge derived from the surface structure of the sentence permits us to perceive the deep structure, or meaning, of the sentence.

In our brief examination of auditory memory and attention, we saw the need to provide for an internal buildup of information of a duration long enough to permit segmentation. Norman (1972) suggests that this precategorical or echoic memory first involves an analogue storage of a very short time segment of acoustic information in the form of a sensory wave. A storage time of several seconds permits the buildup necessary for the extraction of suprasegmental and segmental information, so that the message can be restructured into linguistic segments.

Since echoic memory is not equipped to hold segmented data, the information is then passed into short-term memory where it is subjected to conscious control. It now assumes the form of meaningfully sized units which can be held in consciousness through rehearsal.

These units are then processed into long-term memory. The apparent limitation imposed by a long-term memory retention capability of 7 ± 2 segments is overcome by grouping the information into much larger sized cognitive "chunks," each containing several segments. The permanence of long-term memory was explained on the basis of changes in the transmission potentials of neural units influencing their interaction with other units.

Our final question was how the auditory perceptual system can attend to certain patterns of information while other simultaneously occurring patterns in the waveform are ignored to varying degrees. Three possible theories were suggested: Broadbent's (1958, 1962) theory of filtered channels, Treisman's (1960) modification of that theory to allow for relative tuning of channels, and Neisser's (1967) concept of analysis-by-synthesis. This last process accounts for selective perception through sharp focusing of only those pattern components which match the internally generated pattern.

In conclusion, the evidence suggests that by virtue of the structure which the listener imposes upon the incoming acoustic signal, he need perform only a cursory examination of that signal. *We perceive according to the probabilities we have used to generate expectancies.* These probabilities are computed on the basis of sampling the acoustic information, yet that very sampling is itself influenced by expectancies derived from earlier structure and past experience. It would seem reasonable to conclude, therefore, that we perceive according to the manner in which we have prepared our perceptual system to approach the task. We perceive not the signal, but how we have processed it. If there are innate deficiencies in the perceptual system which interfere with, or impede, the development of appropriate perceptual postures, then perceptual behavior will deviate from the norm. When the impaired perceptual function is auditory and involves perception in the speech mode, the symptoms may manifest themselves as difficulties in receiving, discriminating, segmenting, storing, or recalling meaningful speech. Because of the intimate relationship between reception and production, difficulties in speech and language production may also occur. Furthermore, since speech is only one of the symptoms of language, these dysfunctions may cross modal boundaries and appear as difficulties in relating the printed word to the spoken word.

Our next task is to examine the evidence pertaining to how normal infants process speech in order that we can better evaluate the nature of the difficulties associated with deviant auditory perception.

REFERENCES

ABRAMS, K. AND T. G. BEVER, 1969. Syntactic structure modifies attention during speech perception and recognition. *Quart. J. of Exp. Psych.*, **21**, 280–90.

BERRY, M. F., 1969. *Language Disorders in Children.* New York: Appleton-Century-Crofts.

BEVER, T. G., R. KIRK, AND J. LACKNER, 1969. An autonomic reflection of syntactic structure. *Neuropsychologia*, **7**, 23–28.

BEVER, T. G., J. R. LACKNER, AND R. KIRK, 1969. The underlying structures of sentences are the primary units of immediate speech processing. *Perception and Psychophysics*, **5**, 225–34.

BEVER, T. G., J. R. LACKNER, AND W. STOLZ, 1969. Transitional probability is not a general mechanism for the segmentation of speech. *J. Exp. Psychol.* **5**, 225–34.

BJORK, E. L. AND A. F. HEALY, April, 1970. Intra-item and extra-item sources of acoustic confusion in short-term memory. In *Communications in Mathematical Psychology.* Rockefeller University Technical Reports.

BOND, Z. S., 1971. Units in speech perception. *Working Papers in Linguistics*, **5**, (9), 111–12. Computer and Information Science Research Center Technical Report Series. OSU – CISRC-TR-71-8. Columbus, Ohio: The Ohio State University.

BROADBENT, D.·E., 1962. Attention and perception. *Scientific American*, **206**, 143–51.

BROADBENT, D. E., AND P. LADEFOGED, 1959. Auditory perception of temporal order. *J. Acoust. Soc. Amer.*, **31**, 1539.

BROADBENT, D. E., 1958. *Perception and Communication*. New York: Pergamon Press.

COLE, R., M. COLTHEART, AND F. ALLARD. Memory of a speaker's voice: Reaction time to same or different-voiced letters. *Quart. J. Exp. Psychol.*, **26**, 1–7 (1974).

CROWDER, R. J. AND J. MORTON, 1969. Precategorical acoustic storage (P.A.S.). *Perception and Psychophysics*, **5**, 365–73.

CROWDER, R., 1972. "Visual and Auditory Memory," in *Language by Ear and by Eye*, eds. J. F. Kavanaugh and I. G. Mattingly. Cambridge: M.I.T. Press.

DAY, R. S., November, 1969 (Haskins Laboratory), 1970a. Temporal order judgments in speech: Are individuals language bound or stimulus bound? Paper presented at the 79th meeting of the Psychonomic Society, St. Louis SR – 21/22, 71–87.

DAY, R. S., April 1970b. Temporal order perception of a reversible phoneme cluster, Paper presented at the 79th meeting of the Acoustical Society of America, Atlantic City, 21–24.

ECCLES, J. C., 1953. *The Neurophysiology of Mind*. London: Oxford Press.

EIMAS, P. D. In press. Developmental studies of speech perception, in *Infant Perception*, eds. L. B. Cohen and P. Salapatek. New York: Academic Press.

ERIKSEN, C. W. AND H. J. JOHNSON, 1964. Storage and decay characteristics of non-attended auditory stimuli. *J. Exp. Psychol.* **63**, 28–36.

ESTES, W. K., 1972. "The Associative Basis of Stimulus Coding," in *Coding Processes in Human Memory*, eds. A. W. Melton and E. Martin. Washington, D.C.: V. H. Winston

ESTES, W. K., April 1970. On the source of acoustic confusions in short-term memory for letter strings. In *Communications in Mathematical Psychology*. Rockefeller University Technical Reports.

FAIR, C. M., 1965. The organization of memory functions in the vertebrate system. *Neurosci. Res. Prog. Bull.*, **3**, 27–62.

FODOR, J. A. AND T. G. BEVER, 1965. The psychological reality of linguistic segments. *J. Verb. Learn. Behav.*, **4**, 414–20.

GARRETT, M., T. BEVER, AND J. FODOR, 1966. The active use of grammar in speech perception. *Perception and Psychophysiology*, **1**, 30–32.

GARVIN, P. L., "Human language and computer language." In *Anals da II Bienalde Ciencia.*—Humanismo Sâo Paulo.

GAVIN, P. L., "A linguagem humana e a linguagem do computador," in *Anais do Simpósio sôbre Ciência e Humanismo*. São Paulo: Fundaçâo Bienal de Sâo Paulo, 1971.

GUTTMAN, N., AND B. JULESZ, 1963. Lower limits of auditory periodicity analysis. *J. Acoust. Soc. Amer.*, **35,** 610.

HIRSH, I. J., 1959. Auditory perception of temporal order. *J. Acoust. Soc. Amer.*, **31,** 759–67.

HOLMES, V. AND K. FORSTER, 1970. Detection of extraneous signals during sentence recognition. *Perception and Psychophysics*, **7,** 297–301.

JOOS, M., 1950. Description of language design. *J. Acoust. Soc. Amer.*, **22,** 701–8.

LA BERG, D., 1972. "Beyond Auditory Coding," in *Language by Ear and Eye*, eds. J. Kavanagh and I. G. Mattingly. Cambridge, Mass.: The M.I.T. Press.

LA BERG, D., 1971a. Effect of type of catch trial upon generalization gradients of reaction time. *J. Exper. Psych.*, **87,** 225–28.

LA BERG, D., 1971b. On the processing of simple visual and auditory stimuli at distinct levels. *Perception and Psychophysics*, **9,** 331–34.

LASHLEY, K. S., 1950. "In Search of the Engram," in *Physiological Mechanisms in Animal Behavior* (Symposium No. 4 Soc. Exp. Biol.) New York: Academic Press.

LASKY, R., A. SYRDAL-LASKY, AND R. KLEIN, 1975. VOT discrimination by four-to-six-months-old infants from Spanish environments. *J. Exp. Child Psych.*, **20,** 215–25.

LEHISTE, I., 1972. "The Units of Speech Perception," in *Speech and Cortical Functioning*, ed. J. H. Gilbert. New York: Academic Press.

LIBERMAN, A. M., I. G. MATTINGLY, AND M. T. TURVEY, 1972. "Language Codes and Memory Codes," in *Coding Processes in Human Memory*, eds. A. W. Melton and E. Martin. Washington, D.C.: V. H. Winston.

LIBERMAN, A. M., 1970. The grammars of speech and language. *Cognitive Psych.*, (© Academic Press, Inc.) **1,** 301–23.

LIEBERMAN, P., 1967. "Intonation and the Syntactic Processing of Speech," in *Models for the Perception of Speech and Visual Form*, ed. W. Wathen-Dunn. Cambridge, Mass.: The M.I.T. Press.

MARTIN, J. G., 1972. Rhythmic hierarchical versus serial structure in speech and other behavior. *Psychol. Rev.*, **73,** 487–509.

MENYUK, P., 1971. *The Acquisition and Development of Language*. Englewood Cliffs, N.J.: Prentice-Hall.

MILLER, G. A., 1956. The magical number 7, ± 2: Some limits on our capacity for processing information. *Psych. Rev.*, **63,** 81–97.

NEISSER, U., 1967. *Cognitive Psychology* (Century Psychology Series). New York: Appleton-Century-Crofts.

NORMAN, D. A., 1972. "The role of memory in the understanding of language," in *Language by Ear and Eye*, eds. J. F. Kavanagh and I. G. Mattingly. Cambridge, Mass.: The M.I.T. Press.

PIERCE, J. R., 1969. Whither speech recognition. *J. Acoust. Soc. Amer.*, **49,** 1049–51.

STEVENS, K. N., 1960. Toward a model for speech recognition. *J. Acoust. Soc. Amer.*, **32,** 47–55.

STREETER, L., 1974. The effects of linguistic experience on phonetic perception. Unpublished doctoral dissertation. New York: Columbia Univ.

TREISMAN, A. M., 1960. Contextual cues in selective listening. *Quart. J. Exp. Psychol.*, **12,** 242–48.

WICKELGREN, W. A., 1969. Context-sensitive coding, associative memory and serial order in speech behavior. *Psych. Rev.*, **1,** 1–15.

7

Infant Speech Perception

PHILIP A. MORSE

From our previous discussion of speech perception it is apparent that the adult listener performs some rather complex decoding of the acoustic signal in order to lay bare its phonetic content (Liberman, Cooper, Shankweiler, and Studdert-Kennedy, 1967). Of primary interest is how the adult comes to acquire this complex code. In this chapter we shall examine the recent research on *infant* speech perception in an effort to find some answers to this problem. We will be concerned both with major research findings and with the exciting frontiers of investigation on which these focus our attention.

Perhaps the most obvious point at which to begin our discussion of speech perception in infants is the nature of the experimental organism. Anyone who has spent time with babies, either in or out of the laboratory, will readily admit that one- to four-month-old infants are far from ideal subjects. Not only are they unable to verbalize what they know about the world, but also they have a number of behaviors (e.g., sleeping, crying, etc.) which compete quite successfully with our efforts to study them.

To date, two major procedures have been developed to extract linguistic knowledge from the human infant:

1. *the heart-rate (HR) habituation/dishabituation paradigm, or model*
   (In this procedure a physiological response reflecting the infant's

attention to a novel event is seen to decrease in magnitude as the novel stimulus becomes "familiar" and to reappear if a second novel stimulus is presented.)

2. *the nonnutritive high-amplitude sucking (HAS) paradigm* (This is an operant conditioning procedure that employs as the response the infant's strong sucking on a pacifier. Changes in the infant's sucking to a familiar vs. a novel stimulus are used to index discrimination.)

The HR habituation/dishabituation paradigm was derived from the finding that when a sound is presented to an infant at a comfortable listening level, the infant's heart rate slows down (decelerates) in response to the sound over the course of the next 0 to 20 secs. This heart-rate deceleration has been shown to be the cardiac counterpart of the general orienting behavior of the individual. Thus, the cardiac orienting response reflects the subject's *attention* to a novel event (Graham and Clifton, 1966). If a

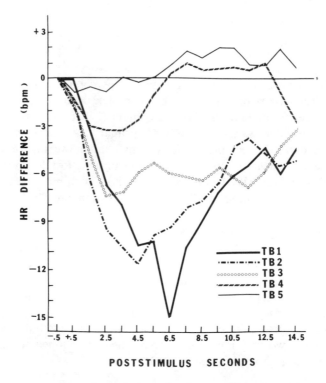

Figure 7.1   Habituation of the orienting response over successive trial blocks (1–5). (From Miller, Morse, and Dorman, 1975)

sound is presented repeatedly to an infant, he will begin to develop a "memory" of its characteristics. This eventually results in the infant's ceasing to respond to the sound as "novel." Consequently, we should expect to see the heart-rate orienting response gradually become smaller and smaller as the stimulus is repeated. This is in fact what we see in Figure 7.1, which shows both the initial orienting response and its habituation over 5 trial blocks. If we change the stimulus after the orienting response of the infant has become habituated to it, we may observe a further orienting response to the changed stimulus. This is the dishabituation explained in (1) above. If this reoccurrence of the orienting response is observed, we can assume that the infant has discriminated the second stimulus from the first one; in other words, he has responded physiologically to the acoustic change manifest in the shift from the first sound to the second one. This orienting response, habituation/dishabituation paradigm has been employed quite successfully in studying the auditory discrimination of infants 4 months of age and older (Berg, 1972; Moffitt, 1971). Its use with infants less than 4 months of age, however, has been rather discouraging (Brown, Leavitt, and Graham, in press; Leavitt, Brown, Morse, and Graham, in press; Miller, Morse, and Dorman, 1975).

The second paradigm for infant testing, the high amplitude sucking procedure, on the other hand, has proven to be extremely useful with infants as young as one to four months of age (Eimas, Siqueland, Juszyck, and Vigorito, 1971; Morse, 1972). The logic of this paradigm is quite similar to that of the heart rate procedure, except that the infant is in full control of the stimulus presentation. As the infant sucks on a nonnutritive nipple, he causes a sound to be transmitted through a speaker. The more the infant sucks on the nipple, the more frequently the sound is presented. If the infant is a cooperative subject, he will eventually learn that the sucking turns on the sound. As the experiment progresses the first thing we observe is an increase in the sucking rate (acquisition). Eventually, as the sound loses its reinforcing properties, the infant's sucking rate will drop off (habituation). When the sucking rate meets some predetermined habituation criterion, the sound is changed. If we observe a recovery of the sucking rate relative to a no-change control group, we can infer that the infant has discriminated between the new and the old stimuli. However, each infant sucks with varying degrees of strength and interest. For this reason, in the beginning of this procedure, the experimenter selects the infant's hardest (high amplitude) sucks (HAS) as those which he will subsequently reinforce. The next step is to allow a one-minute silent period to permit the determination of an average, or a baseline, count of the infant's HAS. After this baseline period only the infant's high amplitude sucks are permitted to cause a sound to be presented. A typical experimental result showing the averaged data for the baseline (BL), for the five minutes prior to stimulus

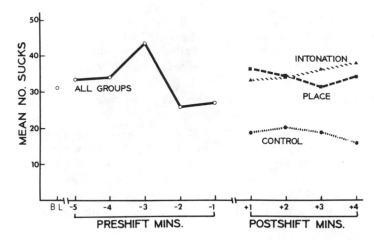

Figure 7.2 Changes in HAS for baseline (BL) five minutes prior
to stimulus shift (-5 to -1), and four postshift
minutes ($+1$ to $+4$). (From Morse, 1972)

shift (-5, -4, -3, -2, -1 mins.), and for the four minutes following
shift ($+1$, $+2$, $+3$, $+4$ mins.), is shown in Figure 7.2. A more extensive
discussion and a comparison of the heart rate and high amplitude sucking
paradigms can be found in an earlier publication (Morse, 1974).

Using these two paradigms, infant speech researchers have begun to
ask several different questions about auditory perception in the human
baby. Some investigators have sought to determine if infants can discriminate auditorily between different pairs of speech sounds. Positive results
from such studies, however, do not necessarily indicate that the infant is
treating these sounds as speech, or processing them in a speech mode. This
problem has stimulated other researchers to try to answer this particular
question; namely: Does the infant utilize the *phonetic* categories of adult
speech in organizing his discriminative behavior, i.e., low level linguistic
processing, or is he simply classifying the sounds as acoustically different?
(Remember, we referred to this latter stage of auditory perception in Chapter 6 as "primary acoustic processing.") As we review the major findings in
infant speech perception, we should bear in mind this important distinction
between auditory and phonetic discrimination, a distinction which has already been stressed.

The first important study in the recent history of infant speech perception asked whether 4- to 5-month-old infants could discriminate two
syllables which differed in *place* of articulation (bilabial, [p/b],[1] vs. velar

---

[1]Elsewhere in this book slashes have been used to describe speech sounds in
terms of *phonemic* classes (for example /ba/). However, since it would be

sounds, [k/g], for example). Moffitt (1971) used a heart-rate habituation/ dishabituation paradigm to show that infants at this age could discriminate auditory [ba] vs. [ga] when the only acoustic difference between the stimuli was in the second formant (F2) transition of each. I have also used the HAS procedure to demonstrate that as early as 6 weeks of age infants can discriminate [ba] vs. [ga] when they differ in both F2 and F3 transitions (Morse, 1972). Furthermore, if infants are presented with the F2 and F3 transitions in isolation (these sound like chirps or glissandos), they discriminate them in a manner different from the way they do when the same transitions are presented in the speech contexts [ba] and [ga]. Most recently, Eimas (1974) confirmed the finding that infants can discriminate acoustic cues which signal place of articulation contrasts. In addition, he showed that they can do this not only auditorily, but also phonetically. When 2- to 3-month-old infants were tested with the high-amplitude sucking paradigm, a change between categories ([dae] vs. [gae]) resulted in a large increase in the HAS rate. However, the infants showed no response recovery for either a no-change control or for a comparable acoustic shift within categories (within [dae] or within [gae]). That is to say, an acoustic shift in the F2 and F3 transitions evoked a change in response when it crossed a phoneme boundary, but a shift of equivalent magnitude within the category failed to do so. Since the acoustic differences in the F2 and F3 transitions were identical both in the between- and in the within-category conditions, these results show that the discriminations infants are able to make are constrained by adult phonetic categories. That is, they discriminate place contrasts only when the *phonetic categories are different*. (A study by Miller and Morse, [in press] has recently replicated this finding of categorical discrimination for place of articulation using a cardiac orienting response paradigm.) Finally, Eimas (1974) has also observed that if the same F2 transitions signaling the distinction [dae] vs. [gae] are presented in isolation (chirps), infants discriminate not only between categories, *but also within categories*. These transitions, in this case, are processed as continuous rather than categorical stimuli. This provides additional evidence that the infant is capable of phonetically discriminating the acoustic cues for place of articulation.

Similar research has been conducted with the acoustic cues that signal the voicing distinction in stop consonants [p/b, t/d, k/g] in the initial position. Such cues permit differentiation, for example, between [ba] and [pa]. These cues, labeled collectively voice-onset-time (VOT), are definitely discriminated auditorily by infants. Trehub and Rabinovitch (1972),

premature to argue that young infants perceive speech sounds according to minimal differences in the meaning of words (phonemically), brackets (e.g. [ba]) have been employed throughout most of this chapter to describe speech sounds as *phonetic* categories.

using the HAS paradigm, observed that 4- to 7-week-old infants could dis-discriminate both natural and synthesized [ba] and [pa] syllables. How-ever, Eimas, Siqueland, Juszysk, and Vigorito (1971) found that by one month of age (and also at four months) the HAS procedure revealed that infants discriminated these voice-onset-time cues between the categories [ba] vs. [pa] but not within these voicing categories. This finding of *phonetic*, categorical discrimination has been recently replicated by Miller (1974) and extended to the voicing contrast in [da] and [ta] (Eimas, 1975b). Thus, these findings for infant VOT discrimination are in complete accord with the perceptual behavior for place cues already discussed.

In contrast to the finding that both adults and infants *categorically* discriminate stop consonants varying in place of articulation and voicing, vowels are generally perceived continuously rather than categorically (Stevens, Liberman, Studdert-Kennedy, and Oehman, 1969; Pisoni, 1971, 1973). Trehub (1973) demonstrated, using the HAS paradigm, that 4- to 17-week-olds could discriminate auditorily the vowel contrasts [i] vs [a] and [i] vs. [u]. Most recently, Swoboda, Morse, and Leavitt (1976) showed that 8-week-olds, like adults, do not discriminate the vowels [i] and [ɪ] categorically. They are able to discriminate equally well between-*and* within-category contrasts. Thus, for both adults and infants the pho-netic categories of vowels (at least vowels 200–300 msec in duration) do not preclude the discrimination of within-category differences, thereby yielding continuous rather than categorical discrimination.

Using the high-amplitude sucking paradigm, additional studies of speech cues have revealed that infants are able to discriminate falling vs. rising intonation in the syllable [ba] (Morse, 1972); categorical differences in the liquids [ra] vs. [la] (Eimas, 1975a); and differences in the fricatives [va] vs. [sa], [sa] vs. [ʃ a], but not in [sa] vs. [za] (Eilers and Minifie, 1975). Finally, Miller, Morse, and Dorman (1975), using a heart-rate paradigm to be described below, found that 3-month-olds could discrimi-nate the very brief burst cue (5 to 30 msec) which occurs at the release of initial stop consonants.

In sum, all of these studies in infant speech perception reveal that infants can discriminate almost every relevant acoustic cue(s) in those speech sounds that have so far been presented to them. Furthermore, research has shown that, as in adults, discrimination between stop con-sonants is phonetic (i.e., categorical), whereas in vowels it resembles continuous (nonphonetically constrained) discrimination. This research thus supports the position that *some aspects of processing in a speech mode are either a genetically endowed capacity in infants, or they are learned within the first few weeks of life*. Furthermore, it indicates that infants, like adults, process vowels and consonants differently (continuous vs. categorical dis-crimination).

The data on infant speech perception suggest, therefore, that at a very early age the infant's auditory and phonetic capabilities are quite sophisticated. The findings raise a number of very important issues about the factors that might influence the development of these impressive capabilities. In the remainder of this chapter we shall explore some of the theoretical implications of these findings as they relate to the development of speech perception in human beings.

The Role of Memory Factors

If we were to present an infant with a [ba] sound on Monday and with a [ga] sound on Wednesday and ask whether he could discriminate them, we should not be too surprised if, when tested, we found that he was unable to do so. Although this example is exaggerated, the point I wish to make is that if we exceed the infant's short-term (STM) or long-term memory (LTM) abilities in our testing procedures, we may fail to determine the infant's true discriminative capacities. Brown, Leavitt, and Graham (in press) recently studied the development of the discrimination of triangular wave stimuli (sounds sweeping four times a second between 1500 and 2500 Hz or between 150 and 250 Hz) using a 6/2 habituation/dishabituation heart-rate paradigm. In this paradigm, 6 habituation trials of the same stimulus (repeated 5 to 6 times per trial) were followed by 2 trials of a novel stimulus. Each trial was separated by 25 to 35 secs of silence. Six-week-old infants failed to show an initial orienting response to these stimuli (trials 1 and 2) and exhibited no habituation or dishabituation on subsequent trials. Nine-week-old infants did orient to the onset of these stimuli, but failed to show either habituation or dishabituation. Finally, in a subsequent study (Brown, personal communication) twelve-week-old infants showed both orienting and habituation, but no dishabituation to the change stimulus (trials 7 and 8). A second study by Leavitt, Brown, Morse, and Graham (in press) also failed to find auditory discrimination (dishabituation) with a 6/2 heart-rate paradigm in 6-week-old infants. However, when a paradigm was used that completely eliminated the silent intervals between trials, 6-week-old infants evidenced differential responding to a similar acoustic change. In this No-intertrial-interval (No-ITI) paradigm a series of one stimulus was followed immediately (No-ITI) by a series of a novel stimulus.

In marked contrast to the failure to demonstrate auditory discrimination in six-week-old infants with a 6/2 paradigm, Moffitt (1971), Berg (1972), and Lasky, Syrdal-Lasky, and Klein (1975) have reported that an habituation/dishabituation paradigm *does* yield successful auditory discrimination in infants four months of age or older. Thus the Leavitt et al. study taken together with the findings of Moffitt, Berg, and Lasky et al. *suggest* that between six weeks and four months of age the short-term memory constraints of the paradigm (i.e., the intervals between the trials) may have important consequences for the assessment of the infant's auditory discriminative abilities. The failure of the younger infants to exhibit auditory discrimination in the 6/2 paradigm may be related to their relatively poor ability to retain the auditory image in short-term memory long enough to permit the necessary discriminative comparison to be made. This is an important consideration if we are to attempt to define the components of auditory perception. Additional evidence for this position derives from the Miller et al. (1975) study of burst discrimination, mentioned earlier. In this study, three-month-olds showed orienting and impressive habituation to the syllables [bu] and [gu], but failed to demonstrate dishabituation even though an 8/2 (8 habituation trials/2 change trials) paradigm was employed. In contrast, infants of the same age *did* demonstrate discrimination of the burst cue in these stimuli when a No-ITI paradigm was employed. In this particular study infants simply heard 20 presentations of one stimulus followed by 20 presentations of the syllable change with no interval between the two. As in the Leavitt, Brown, Morse, and Graham study (in press), heart-rate change was measured at the stimulus shift.

Memory factors in infants affect not only whether discrimination occurs but also the quality of that discrimination. Pisoni (1971, 1973) and Fujisaki and Kawashima (1969) have both suggested that an auditory short-term memory (STM) and a phonetic STM component play an important role in speech discrimination. If the stimuli or the paradigm is such as to permit the subject to retain *an auditory short-term memory* of the fine details of the stimulus, then discrimination within categories will be enhanced. In contrast, a paradigm, which forces the subject to reply primarily on his *phonetic short-term memory* of the stimuli, will make within category discrimination more difficult. Specifically, the extent to which vowels are perceived categorically is influenced by the duration of the stimuli and by the intervals between stimuli in the discrimination task (Pisoni, 1971, 1973). Long vowels tend to be perceived more continuously than short vowels (for which within-category discrimination is reduced). Following this logic, we could ask about the human infant: Does the extent to which auditory short-term memory is available for vowels (short vs. long stimuli) affect the categorical-continuous nature of the infant's discrimination? As was indicated earlier, Swoboda, Morse, and Leavitt (1976) found that when the stimuli had a 240 msec duration, 2-month-old infants discriminated [i] and [ɪ] continuously. If the duration of the stimuli is reduced to 60 msec, reliable between-category discrimination remains but within-category discrimination is not significant. The difference between the between- and within-category conditions, also is not statistically significant (Swoboda, Kass, Morse, and Leavitt, in preparation). The failure of these two conditions to be statistically significant from each other is not surprising since Pisoni observed that shortening the duration of vowels for adults made perception relatively more categorical, but not as impressively categorical as observed for consonants. These results suggest that the infant's discrimination of vowels, like the adult's, can be made more categorical or more continuous as a function of stimulus duration. Presumably, this effect is related to the extent to which auditory short-term memory for vowels is available to the infant. Reducing the stimulus duration apparently compels infants (and adults) to rely more on their categorical, phonetic short-term memory of the stimuli, thus yielding more categorical-like discrimination.

How "Phonetic" Is "Phonetic?"

In reviewing the literature on infant speech perception, we stressed at the outset the difference between primary acoustic discrimination and linguistic discrimination. Some studies examined the infant's ability to discriminate *auditorily* various speech contrasts (e.g., Moffitt, 1971), whereas other studies have investigated the *phonetic* quality of this discrimination (Eimas, 1974). However, those studies which have examined the phonetic capabilities of infants have done so within a very restricted definition of "phonetic." In the case of stop consonants, they have looked at categorical discrimination of a particular acoustic cue(s) under conditions in which the vowel is held constant (Eimas, 1974; Eimas et al., 1971). We have already seen, in our examination of studies on adult speech perception, that the acoustic cues for consonants vary as a function of vowel context and position in the word. Figure 7.3 illustrates, for example, that the critical cue for the place of articulation of the alveolar dental consonant [d] may be radically different in [di] and [du]. Similarly, the bilabial consonant "p" may be cued by voice-onset-time cues as in "pit" vs. "bit," by silence as in "split" vs. "slit," or by vowel duration as in "sip" vs. "sib." In other words, the "phonetic" category "p" is much more than the voice-onset-time category examined in studies of categorical perception. It is quite plausible that these more "abstract" phonetic categories do not emerge in linguistic development until much later than the categorical discrimination observed in 1- to 3-month-old babies.

To date, we know very little of the infant's ability to organize these more abstract phonetic classes. This is primarily due to the fact that both the heart-rate and sucking paradigms are designed to study discrimination. What is needed is a paradigm which contains two different responses and permits us to study the infant's generalizations or identifications *among*

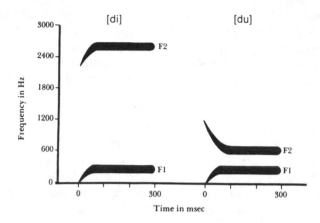

Figure 7.3 A simplified acoustic representation of the syllables [di] and [du]. (From Mattingly, 1972, © The Society of the Sigma Xi, New Haven, Conn.)

stimuli. Recently, Fodor, Garrett, and Brill (1975) have reported the success of a head-turning paradigm in studying this very problem. They examined the infant's learning of a head-turning response to natural speech syllables such as [pi], [ka], and [pu]. Their results indicate that 4-month-olds find it easier to associate syllables (learn to turn their heads to the same side) that have the *same* (e.g., [pi] and [pu]) rather than different (e.g., [pi] and [ka]) initial consonants. If we assume there were no invariant *acoustic* cues in the natural [pi] and [pu] stimuli used, cues which might have aided infants in learning this task, then these findings do provide some indication that infants are able to classify very different acoustic stimuli into the same abstract, phonetic class. Obviously, much more research with synthetic and natural speech stimuli is necessary before we fully understand the infant's acquisition of the many levels of "phonetic" classification.

Is Speech Special?

If infants do have the ability to discriminate speech sounds in an adult-like manner at such an early age, they may also be equipped to process speech in a special mode; this may be a consequence of the evolution of uniquely human capabilities for speech (Lieberman, 1974). Such a special mode of processing was considered in Chapter 5. Recent studies have begun to examine this problem in infants in three different ways:

1. hemispheric differences for speech vs. nonspeech

2. the categorical discrimination of speech vs. nonspeech
3. the attention (orienting) that the infant pays to speech and non-speech stimuli.

Before we explore some of these findings, a word of caution is in order regarding the differences between speech and nonspeech stimuli. It is virtually impossible to produce a nonspeech control stimulus perfectly equated (acoustically) with a speech stimulus which is *not* perceived by listeners as speech. You will recall from earlier discussions that the auditory system must make an early decision concerning whether a signal is to be analyzed as speech or nonspeech. Providing a stimulus meets the minimal criteria for classification as speech, it will be processed and perceived as speech regardless of whether or not it is speech. Consequently, all of the studies to be discussed must be viewed as exploring differences between responses to particular types of nonspeech and speech stimuli.

Studies dealing with the specialized role of the two hemispheres of the brain suggest that young infants may perceive speech better in the left hemisphere and nonspeech better in the right hemisphere. Molfese, Freeman, and Palermo (1975) observed that when infants averaging approximately 5 months of age were exposed to speech stimuli ([ba], [dae], "boy," "dog"), they exhibited the strongest cortical responses over the left hemisphere, whereas the greatest responses to nonspeech (C major chord, 250 Hz to 4kHz noise burst) occurred over the right hemisphere. Entus (1975) has recently reported related findings using the high-amplitude sucking procedure with the two ears receiving competing stimuli. She demonstrated that infants between 2 and 4 months of age discriminated a speech change better in the right ear than in the left ear, whereas a nonspeech change was discriminated better in the left ear. Although both of these findings need to be replicated, they do suggest that at a very early age the infant's brain may be responding to speech sounds in a special (lateralized) way.

Research on the discrimination of speech vs. nonspeech stimuli has shown that the transitions of the second formant (F2) in a speech context ([ba/da]) are discriminated categorically by infants 2 to 3 months of age, whereas the F2 transitions presented in isolation are discriminated *continuously* (Eimas, 1974). On the other hand, a recent study by Juszyck, Rosner, Cutting, Foard, and Smith (1975) has revealed that nonspeech stimuli varying along a continuum of rise-time (that is, stimuli with varying rates of onset buildup) are discriminated *categorically* (as is the case for adults). *These findings suggest, at the very least, that young infants can discriminate speech and nonspeech stimuli in a manner similar to adult listeners.* Whether the infant (or adult) necessarily discriminates all nonspeech stimuli differently from speech stimuli remains to be determined.

Finally, Leavitt, Brown, Morse, and Graham (in press) observed that 6-week-old infants pay attention (orient) to *pulsed* (with silent interval between stimuli) speech ([ba], [ga]) and nonspeech (sine wave) stimuli. However, when these same speech stimuli were presented nonpulsed (continuously, with no silent interval between stimuli), infants continued to pay attention to them (Morse, Leavitt, Brown, and Graham, in press) but failed to orient to the nonspeech triangular wave stimuli of the Brown, Leavitt, and Graham (in press) study. Although these results may not necessarily be valid for other types of nonspeech stimuli, they do suggest that under some conditions speech stimuli may be attended to more than other auditory stimuli. In sum, these three types of research offer some *suggestive* evidence that under some conditions infants respond to speech stimuli in a "special" manner.

Role of Experimental and Innate Factors

Certainly one of the most interesting questions posed by this research is: How does the infant acquire phonetic categories in the development of speech perception? Of particular interest is the fact that the development of these perceptual categories precedes the infant's production of these phonetic contrasts (e.g., Kewley-Port and Preston, 1974). Although we do not yet know exactly to which sounds the infant is exposed in the first few months of life, the development of these categories does not appear to depend entirely on the infant's experience with his parents' language. This is suggested by the work of Streeter (1976), Eimas, 1975b), and Lasky, Syrdal-Lasky, and Klein (1975), who have shown that during the first few months of life infants can discriminate differences in the voicing of languages foreign to their home environments. These cross-language findings suggest that perhaps these voicing categories are not acquired ontogenetically (i.e., within the developmental stages of the individual infant), but phylogenetically (i.e., within the development of the human species). Since the production of the full range of speech sounds is unique to the human species (Lieberman, 1974), it may be that human beings have evolved special auditory perceptual capabilities that are optimally correlated with their categories of speech sound production. Behavioral research with adult subjects using an adaptation paradigm (e.g., Eimas and Corbit, 1973; Cooper, 1975; Diehl, 1975) has suggested that special auditory, and perhaps even phonetic, feature detectors may be operative in the human auditory system (see detailed discussion in Chapter 5). Eimas (1975b) has further proposed that these feature-detector mechanisms, together with similar adaptation processes, may be responsible for the infant phonetic discrimination data discussed above. However, recent discrimination findings in the rhesus monkey (Morse and Snowdon, 1975; Waters and Wilson, 1976) suggest that categorical discrimination of stop consonants may not be limited to human adults or infants. If further research with nonhuman species continues to yield similar results, then we may wish to conclude that, rather than being an ability limited to the human species, some of these discriminative capabilities may be a general property of the primate or mammalian auditory system. Perhaps extending to infants and other species the use of the adaptation procedure employed in human adults may shed more light on the viability of a feature-detector explanation for the ontogenetic and phylogenetic development of some "phonetic" categories in speech perception (Morse, in press).

Individual Differences

Our overview of the literature on infant speech perception suggests that infants with "normal" medical histories are able to grasp many of the important basics of the human speech code at a very early age. The language development literature, on the other hand, indicates that premature infants or those who have experienced a variety of stresses during pregnancy or delivery have a greater chance of later developing problems in language acquisition (Braine, Heimer, Wortis, and Freedman, 1966; Ehrlich, Shapiro, Kimbal, and Huttner, 1973). Consequently, an interesting question is whether or not infants with stressful neonatal histories already exhibit language development problems in their early speech perception capabilities. To date, very little is known about the speech perception of this population of infants. One study, perhaps the only one in this area to date (Swoboda, Morose, and Leavitt, 1976) obtained some suggestive evidence of differences in the within-category vowel discrimination of normal vs. "at-risk" infants. Although the Swoboda et al. study represents an important first step in studying infants at risk for later language development problems, we have only begun to scratch the surface in terms of our knowledge of speech perception difficulties in "at-risk" infants. We must consider not only the wide range of auditory discriminations that the normal infant can make, but also the importance of memory factors, innate and experiential factors, the scope of the "special" nature of speech perception, and the "phonetic" complexities of the speech code. Only after we have assessed these factors, can we begin to comprehend the multitude of ways in which "at-risk" infants may reveal deficiencies in infant speech perception. Hopefully, in the development of the organization of the speech code in the "normal" infant, we shall move closer to detecting problems in the *very* early language development of infants who are at risk for later developmental problems. Such research will be of paramount importance to those who seek not only to understand the complexities of auditory problems in language processing but also to develop effective strategies for early detection and early intervention.

REFERENCES

BERG, W. K., 1972. Habituation and dishabituation of cardiac responses in 4-month-old alert infants. *J. Exp. Child Psychol.*, **14**, 92–107.

BRAINE, M. D. S., C. B. HEIMER, H. WORTIS, AND A. M. FREEDMAN, 1966. Factors associated with impairment of the early development of prematures. *Monogr. Soc. Res. Child Dev.*, **31**, (7).

BROWN, J. W., L. A. LEAVITT, AND F. K. GRAHAM. Response to auditory stimuli in six and nine week old human infants. *Develop. Psychobiology*, in press.

COOPER, W. E., 1975. "Selective Adaptation to Speech," in *Cognitive Theory*, Vol. 1, eds. F. Restle, R. Shiffrin, N. Castellan, B. Landman, and D. Pisoni. Potomac, Md.: Erlbaum, pp. 23–54.

DIEHL, R. H., 1975. The effect of selective adaptation on the identification of speech sounds. *Perception and Psychophysics*, **17**, 48–52.

EHRLICH, C. H., E. SHAPIRO, B. D. KIMBAL, AND M. HUTTNER. Communication skills in five-year-old children with high-risk neonatal histories. *J. Sp. Hrg. Res.*, **16**, 524–29.

EILERS, R. AND F. MINIFIE, 1975. Fricative discrimination in early infancy. *J. Sp. Hrg. Res.*, **18**, 158–67.

EIMAS, P. D., 1974. Auditory and linguistic processing of cues for place of articulation by infants. *Perception and Psychophysics*, **16**, 513–21.

EIMAS, P. D., 1975a. Auditory and phonetic coding of the cues for speech: Discrimination of the [r-l] distinction by young infants. *Perception and Psychophysics*, **18**, 341–47.

EIMAS, P. D., 1975b. "Developmental Studies of Speech Perception," in *Infant Perception*, Vol. II, eds. L. B. Cohen and P. Salapatek. New York: Academic Press, pp. 193–231.

EIMAS, P. D., AND J. CORBIT, 1973. Selective adaptation of linguistic feature detectors. *Cognitive Psych.*, **4**, 90–109.

EIMAS, P. D., E. R. SIQUELAND, P. JUSZYCK, AND J. VIGORITO, 1971. Speech perception in infants. *Science*, **171**, 303–6.

ENTUS, A. K., April, 1975. Hemispheric asymmetry in processing of dichotically presented speech and nonspeech sounds by infants. Paper presented at the meetings of the Soc. for Res. in Child Dev., Denver, Colo.

FODOR, J. A., M. F. GARRETT, AND S. L. BRILL, 1975. Pi ka pu: The perception of speech sounds by prelinguistic infants. *Perception and Psychophysics*, **18**, 74–78.

FUJISAKI, H. AND T. KAWASHIMA, 1969. On the modes and mechanisms of speech perception. *Ann. Rep. Engin. Res. Inst.*, Faculty of Engineering, Univ. of Tokyo, **28**, 67–73.

GRAHAM, F. K. AND R. K. CLIFTON, 1966. Heart-rate change as a component of the orienting response. *Psychol. Bull.*, **65**, 305–20.

JUSZYCK, P., B. ROSNER, J. CUTTING, C. FOARD, AND L. SMITH, April, 1975. Categorical perception of nonspeech sounds in the two-month-old infant. Paper presented at the meetings of the Soc. for Res. in Child Dev., Denver, Colo.

KEWLEY-PORT, D. AND M. S. PRESTON, 1974. Early apical stop production: a voice onset time analysis. *J. Phonetics*, **2**, 195–210.

LASKY, R., A. SYRDAL-LASKY, AND R. KLEIN, 1975. VOT discrimination by four- to six-month-old infants from Spanish environments. *J. Exp. Child. Psych.*, **20**, 215–25.

LEAVITT, L. A., J. W. BROWN, P. A. MORSE, AND F. K. GRAHAM. Cardiac orienting and auditory discrimination in six-week infants. *Develop. Psychol.*, in press.

LIBERMAN, A. M., F. S. COOPER, D. SHANKWEILER, AND M. STUDDERT-KENNEDY, 1967. Perception of the speech code. *Psychol. Rev.*, **74**, 431–61.

LIEBERMAN, P., 1974. On the evolution of language: a unified view. *Cognition*, **3**, 59–94.

MATTINGLY, I. G., 1972. Speech signs and sign stimuli. *American Scientist*, **60**, 327–37.

MILLER, C. L., P. A. MORSE, AND M. F. DORMAN, 1975. Infant speech perception, memory, and the cardiac orienting response. Paper presented at Society for Research in Child Development, Denver, Colo.

MILLER, C. L., AND P. A. MORSE. The "heart" of categorical speech discrimination in young infants. *J. Sp. Hrg. Res.*, in press.

MILLER, J., 1974. Phonetic determination of infant speech perception. Unpublished doctoral dissertation, University of Minnesota, Minneapolis, Minn.

MOFFITT, A. R., 1971. Consonant cue perception by twenty- to twenty-four-week-old infants. *Child Dev.*, **42**, 717–31.

MOLFESE, D. L., R. B. FREEMAN, AND D. S. PALERMO, 1975. The ontogeny of brain lateralization for speech and nonspeech stimuli. *Brain and Language*, **2**, 356–68.

MORSE, P. A. Speech perception in the human infant and rhesus monkey. Proc. of "Conference on Origins and Evolution of Language and Speech," *Annals of the New York Academy of Sciences*, in press.

MORSE, P. A., L. A. LEAVITT, J. W. BROWN, AND F. K. GRAHAM. Discrimination of continuous speech in six-week-old infants, in preparation.

MORSE, P. A., 1974. Infant speech perception: A preliminary model and review of the literature, in *Language Perspectives—Acquisition, Retardation, and Intervention*, eds. R. Schiefelbusch and L. Lloyd. Baltimore: University Park Press, 19–53.

MORSE, P A., 1972. The discrimination of speech and nonspeech stimuli in early infancy. *J. Exp. Child Psychol.*, (© Academic Press, Inc.) **14**, 477–92.

MORSE, P. A. AND C. T. SNOWDON, 1975. An investigation of categorical speech discrimination by Rhesus monkeys. *Perception and Psychophysics*, **17**, 9–16.

PISONI, D. B., 1973. Auditory and phonetic memory codes in the discrimination of consonants and vowels. *Perception and Psychophysics*, **13**, 253–60.

PISONI, D. B., 1971. On the nature of categorical perception of speech sounds. Unpublished doctoral dissertation. University of Michigan, Ann Arbor, Mich.

STEVENS, K. N., A. M. LIBERMAN, M. STUDDERT-KENNEDY, AND S. E. G. OEHMAN, 1969. Cross-language study of vowel perception. *Lang. Speech*, **12**, 1–23.

STREETER, L., 1976. Language perception of 2-month-old infants shows effects of both innate mechanisms and experience. *Nature*, 259, 39–41.

SWOBODA, P. J., J. KASS, P. A. MORSE, AND L. A. LEAVITT. Memory factors in infant vowel discrimination, in preparation.

SWOBODA, P. J., P. A. MORSE, AND L. A. LEAVITT, 1976. Continuous vowel discrimination in normal at risk infants. *Child Development, 47*, 459–65.

TREHUB, S. E., 1973. Infant's sensitivity to vowel and tonal contrasts. *Develop. Psych.*, **9**, 91–96.

TREHUB, S. E. AND M. S. RABINOVITCH, 1972. Auditory-linguistic sensitivity in early infancy. *Develop. Psychol.*, **6**, 74–77.

WATERS, R. S. AND W. A. WILSON, JR., 1976. Speech perception by Rhesus monkeys: The voice distinction in synthesized labial and velar stop consonants. *Perception and Psychophysics*, **19**, 285–289.

8

Defining Auditory Processing Problems

So far we have been concerned with explaining speech perception in terms of pattern processing. Consequently, we have examined the patterning of the acoustic signal and the probable auditory processes involved in reception, identification, and interpretation of the pattern information. We saw that the final percept is arrived at by integrating various types of information in accordance with the probabilities generated by the language system. We have treated the process of speech perception as a highly dynamic one, dependent upon an auditory system involved in establishing relationships between the internally generated expectancies and the received acoustic signals. It is impossible to appreciate the amazing efficiency with which this system performs such a highly complex activity. We take for granted our skill in processing spoken language at high speed, although for the most part we are unaware of how we do so.

For most of us the system functions smoothly from birth to old age. Some children, however, present evidence to indicate that at one level or another the process is not operating satisfactorily: The child is clearly unable to cope with certain demands placed upon him by the environment. The effects of a dysfunction in the auditory processing of speech manifest themselves at an early age by interfering with normal speech and language acquisition. On the other hand, symptoms may fail to appear until the

177

more sophisticated aspects of processing are required by the task of learning special skills—i.e., phonics, reading, or attending to specific oral-content information. The effects may be so slight as to appear only when second-language learning is attempted. Unfortunately, the subtler problems of auditory processing are frequently not identified as such. Often, a child's deviant behavior is misinterpreted, causing him to be considered intellectually slow, emotionally disturbed, or even hard of hearing. The child's reaction to his difficulties may give rise to patterns of behavior which are merely symptomatic of the problem, leading to his being labeled as a daydreamer, as obstinate, disinterested in learning, or willful and aggressive. This in turn may result in inappropriate management of the child by parents and teachers who experience puzzlement and frustration arising from their unsuccessful attempts to deal with the child's behavior.

Disorders of auditory language processing are often, therefore, of an insidious nature. Only by assessing the child's capacity to handle the several aspects of auditory processing can we hope to obtain a clearer understanding of the nature of the difficulties he or she is encountering. Definition of the problem is a prerequisite to structuring the learning situation. Our goal must be to facilitate the development of new perceptual strategies to surmount the specific processing difficulties experienced.

Using the Theoretical Model for Problem Analysis

The knowledge and perspective of auditory perception we have hopefully gained now become our tools in analyzing the problems of auditory and visual language processing. If our theories possess validity, we should be able to come a little closer to understanding these problems. In other words, we should be able to approach a rationale for an observed dysfunction of the perceptual systems. I refer to a disruption of perceptual *systems* rather than of the auditory perceptual system alone, since it is my firm conviction that the organism functions in an integrative manner. As Thorne has stated:

> All behavior reflects the integrative status of the whole person which simultaneously senses, perceives, learns, retains, feels, thinks and acts as a global unit. The attempt to deal with functions separately is a logical and semantic artifact. (1967, p. 8)

The auditory perceptual model we have examined emphasized that the sensori-perceptual systems do not function as independent processors of sense-specific sensations. Instead, they are involved in the analysis and interpretation of information compatible across sense modalities through the medium of pattern equivalency. Furthermore they are correlated to motor behaviour. The appreciation of this concept is important to an understanding of the interaction of sensori-perceptual systems in such tasks as reading (auditory/visual), writing (auditory/visual/motor), and speaking (auditory/motor/kinesthetic). You will recall that both the active and passive theories of speech perception recognize the role that neuro-motor information plays in restructuring speech signals for comprehension. The two schools of thought differ only in their assessment of the extent of that role.

In view of the complexity of the auditory perceptual task, it is understandable that interference with any stage of the process is likely to result either in an inability to arrive at an interpretation of the pattern information or in the incorrect identification of that pattern. The problems of the learning-disabled child, therefore, extend beyond the relationships between auditory and visual modes of verbal language processing. Ayers (1975) has pointed out that perception and purposeful motor movements, which include those of speech articulation, reading and writing, are largely dependent upon the integration of multi-sensory information. She maintains that the ability of the human organism to produce and integrate sensations through movement is of equal importance to the ability to receive sensations. This view concurs with the active mediated theories of perceptual learning already considered.

Ayers emphasizes that perceptions and learning are not dependent upon simple cortical interpretation of auditory-visual input stimuli but require an hierarchy of previous processes. Most of this earlier processing involves sensory integration and cross-modal (intersensory) correlations. You will recall that we considered evidence supportive of this view (Chapter 3). Ayers goes further to argue that auditory and visual perceptual skills essential to academic learning do not develop in isolation, but are dependent upon the normal integrative functioning of other sensory systems, notably proprioception. Thus in considering learning disabilities we must realize that it is highly probable that the problems experienced in a sensori-perceptual function are closely related, not only to the status of that and other sensory modalities, but to motoric perception. A learning disability, within the framework of Ayer's theory, must be considered at least potentially sensorimotor in nature. She states quite categorically (1975, p. 316) that remedial approaches to learning disabilities should not separate sensation and movement.

Applied to language processing, these concepts contribute to our understanding of the role which the productive aspects of speech may play in a child's ability or disability in perceiving and articulating the sounds of speech. Moreover, the nature of perceptions cannot be observed, they can only be *inferred*. Thus, in assessing verbal learning problems we are dependent on motoric responses in terms of speech or writing. It is from these forms of motor behavior that we make inferences about perceptual difficulties. However, if these two phases of processing are as closely interwoven as the evidence suggests, then the determination of the true nature of the observed deviancy will prove to be a very complex task.

Peripheral vs. Central Processing Problems

Our understanding of the perceptual system as encompassing those physiological organs and tissues from the receptive mechanism of the cochlea to the auditory cortex has stressed a coordinated functioning of all components at all levels. This is made possible in part by the tuning of this self-adjusting servosystem through the mechanism of feedback provided by the efferent fibers. Many students will argue that we should exclude a discussion of peripheral hearing difficulties on the grounds that the peripheral system is concerned with sensation, not perception. We disagree. However, to enter into a discussion of the relationship between sensation and perception would prove nonproductive and confusing. We can avoid the problem by accepting the definition of auditory perception offered by Barr. This most appropriately describes the view compatible with our model:

> . . . receiving, processing, and classifying sensory information for coding into familiar symbols. (1972, p. 3)

There is, however, a fundamental difference between the actual reconstruction of an analogue of the acoustic event and the processing of the information which is derived from that highly encoded pattern. The peripheral process, involving the internal representation of the raw auditory information in its precategorical form, is dependent upon sensitivity to the full range of energy vibrations present in the acoustic signal. The subsequent processing and analysis of the internalized pattern, on the other hand, is dependent upon the ability of the system to determine the spatiotemporal interrelationships of the component information.

Auditory processing difficulties arising at post-cochlear levels differ from cochlear problems because, as Reagan points out, they arise not from a lack of stimulation but from

. . . an inappropriate selection by the organism for developmental
sequencing for excitation. (1973, p. 7)

Post-cochlear processing involves, therefore, the integrative sequencing
processes necessary for the correct reconstruction of the patterns of auditory information.

Despite this difference in the nature of the two types of processing
problems, we must begin our discussion of auditory perceptual difficulties
with a brief consideration of congenital hearing deficiency. Not only does
our model include this level, but behavioral characteristics arising from
either of the two conditions are not always immediately and unequivocally
identifiable.

It is difficult to identify the cause of auditory processing disorders
because different etiological factors are often characterized by many
of the same behavioral symptoms. (NNDS Monograph, No. 9, 1969,
p. 17)

Peripheral Auditory Processing Problems (Hearing Deficiency). Although the child with a hearing deficiency which reduces his *sensitivity*
to sound patterns experiences different types of difficulties from the child
with higher level processing problems, both have trouble in perceiving
the pattern correctly. The congenitally hearing-impaired child, deprived
of much of the information embedded in, or derived from, the acoustic
signal, must ultimately experience an auditory perceptual figure different
from that of a child who receives and regenerates the full range of information.

A deficit in hearing acuity which blocks incoming acoustic stimuli
will interfere with the perceptual process. . . . (Nat. Inst. Health,
1969, p. 9)

To assume that the problem can be explained purely on the basis
of the pattern distortion occurring at the organ of Corti is to ignore the
fact that auditory perception is a developmental process. It reflects an
ontogenetic evolution stimulated by the quality of interaction between the
organism and its environment. If a peripheral auditory deficit precludes
normal interaction, it is logical to assume that the resultant auditory perceptual world will be different, even though equivalency of values may
be achieved with training. When normal levels of communicative behavior
are attained, one must assume that they are based upon compensatory
strategies for information processing.

Experience in teaching speech and language to hearing-impaired children with moderately severe and severe sensory deficits has proven that, if auditory stimulation is provided from a very early age, the potential for the development of intelligible speech is great. If, however, the child is not exposed to amplified sound and appropriate auditory stimulation during the early years of life, the potential for learning these skills is considerably less. The potential deteriorates as the child grows older (Fry and Whetnall, 1954; Wedenberg, 1961; Lenneberg, 1967; Horton, 1974; Lloyd, 1976). The innate biological readiness for the development of a particular system must be triggered by appropriate stimulation at or around the optimal time period. If this stimulation is delayed long enough, the natural plasticity of the organism is lost, seriously impeding, or even preventing, the development of skills dependent upon the sensori-perceptual system involved. Oakland and Williams have emphasized the importance of this environmental triggering and molding. They state:

> . . . auditory abilities do not evolve naturally. They are learned—
> they are shaped largely by prevailing environmental factors. (1971,
> p. 12)

Barsch (1967) has even commented that the listening competency of normal-hearing children is dependent upon the demands placed upon them during the developmental years. The hearing-impaired child, can, therefore, be expected to experience difficulty in subconsciously acquiring speech processing rules simply because the amount of raw internalized acoustic data is sharply reduced. Most of the information we use in the perception of speech is not immediately available *in* the acoustic signal, but must be computed *from* it over a period of time. The hearing-impaired child suffers not only from a reduction in the constraints received (information), but also from the reduced potential for generating information internally. As a result, he frequently fails to understand what is said to him even though he hears the person speaking, or he may misperceive what is said, confusing sound, word, or sentence patterns. Since the process of learning to read is heavily dependent on the ability to link auditory/visual patterns to auditory/visual/motor (speech) patterns, this ability is also usually impaired. The hearing-impaired child characteristically reads at a level below his normal-hearing peers. Furthermore, his abilities to deal with abstract ideas, to pay attention in a normal classroom, to discriminate against a noise background or against a competing message signal, are likely to be affected.

Each of these behavioral symptoms is also characteristic of the child with auditory processing difficulties of a central rather than a peripheral nature. The difference between the problems of the hearing-impaired child and the child with central processing difficulties is primarily that, if given sufficient auditory information early enough, the hearing-impaired child has the capacity to process it. *The system has the potential for normal integrative function.* The child with central processing difficulties, on the other hand, does not lack the raw data, he lacks the ability to analyze it in the manner necessary for its complete and correct interpretation.

As Ayers has emphasized:

the sensori-motor problems of learning disabled children do not lie in input but in internal coordination or processing of that input. The output or motor aspect is a problem only because it is dependent upon the processing of the input. (1975, pp. 301–302)

Kirk (1963) has pointed out that the methods of managing and training the child with sensory deficits differ from those demanded by the problems of the child with so called "learning disabilities." We cannot, for this reason, peremptorily dismiss peripheral hearing disabilities as being beyond the scope of our topic. It is crucial that the first aim of assessment and diagnosis of auditory processing problems shall be to differentiate between peripheral and central auditory processing dysfunction.

A complete and thorough auditory evaluation by an experienced audiologist should precede the assumption that a child has a central processing difficulty rather than a peripheral-receptive one, or vice versa. It is just as damaging to misinterpret the problems experienced by a hearing-impaired child as it is to misinterpret those experienced by a child with central auditory processing difficulties. Chelfant and Scheffelin have stated:

Both central and peripheral disorders can have a disrupting effect on language acquisition. For this reason it is necessary to more clearly define those behaviors which are associated with central and peripheral disorders, and develop procedures to differentiate between the two conditions. (1969, p. 18)

Johnson and Myklebust also stress the dangers in making a hasty decision concerning the nature of the auditory processing difficulty.

. . . the limits within which sensory abilities are assumed to be adequate, i.e., *not* to cause a detriment to learning, must be established as rigorously as possible if a definition is to be applied to those who have learning disabilities as a result of a dysfunction in the brain. Only then can we differentiate between the learning disabilities that are neurological in origin and those that are caused by sensory deprivation. (1967, p. 11)

Evidence of the need for such audiological assessment has been provided by Katz and Ilmer (1972, p. 554), who state that in one group of
"learning disabled" children aged 6 to 14 years, 33 percent were found
to have peripheral hearing losses compared to an expected figure of 1 to
3 percent.

We will not discuss the various diagnostic procedures used in the
evaluation of peripheral hearing deficiency. These procedures and the
special educational measures used to help children compensate for their
auditory learning difficulties are treated thoroughly elsewhere. The assessment of peripheral auditory function has become quite a sophisticated
process and falls within the province of the audiologist, who works closely
with the special teacher of the hearing-impaired or the hearing therapist.
The point is that *there is no justification in assuming that auditory processing problems of an integrative nature exist before the peripheral receptive stage has been shown to be intact.* This is critically important, since
most children with peripheral hearing deficiency derive great benefit from
the use of a hearing aid which makes available more of the raw acoustic
data. Equally important is the fact that amplification maximizes the influence of auditory stimulation on the developmental processes of the auditory perceptual system. These children require a specialized educational
approach from the earliest possible age, since their greatest hope comes
from early intervention. The maximum benefit period has been shown to
occur before the child is two years old. During this period, training to
ensure optimal listening behavior, i.e., the development of an auditory
mode of information processing, is highly profitable.

Central Auditory Processing Problems

Dysfunctions arising from an impairment of the processing function
present a different problem. This category of problems is, unfortunately,
a very confusing one, since the labeling and classification procedure leave
much to be desired. Reading the literature on the topic is a most frustrating
experience because of the many overlapping subcategories. Some components are included in a particular section by one writer only to be excluded
by another. To confound the situation, behavioral characteristics associated with a particular condition are seen to be present in one or more
other conditions. However, since all of these problems affect learning
ability, this common factor has come to provide the umbrella under which
children with central nervous system dysfunction have been gathered.

We have already used the term "learning disability" in this chapter.
It is a term that appears to be quite straightforward in its implications.

However, it has come to be used as a diagnostic label and thus deserves some discussion.

Learning Disabilities. The term "learning-disabled" has replaced the far more misleading label of "minimal brain damage." This term was developed to soften the even more negative aspects of the broader label "brain damage." The term "minimal brain damaged" was initially used to identify children who evidenced problems in processing information despite normal sensory end-organs and intact mental and motor function. The label was clearly intended to imply the presence of a dysfunction of the brain arising from deficit or damage insufficiently severe to manifest gross neurological symptoms.

Johnson and Myklebust (1967), among others, have pointed out that brain dysfunction resulting in perceptual disturbances does not necessarily arise from damage. Thus, the use of the term "minimal" only further complicates the issue (Mylkebust, 1964; Birch, 1964). In 1966, Cruickshank edited a collection of papers on the topic of brain-injured children. The papers had been presented at a conference of twenty-seven experts who rigorously debated the opinions expressed. In discussing the deliberations of that conference, Hallahan and Cruickshank (1973) stated that the participants were unable to reach a decision on the critical issue of nomenclature. There was, however, a general consensus among the educators and others that an educationally relevant behavioral definition, rather than a clinical one, should be adopted, although none has since been formulated (Hallahan and Cruickshank, p. 71). Hallahan and Cruickshank emphasized the trend of professional opinion, when they questioned whether the same group of people would have defended so firmly the need for a non-etiological definition five years earlier.

Johnson and Myklebust (1967) support the need to focus attention on the behavioral manifestations of the problem rather than on the degree of involvement in the brain. They point out that it is the learning processes which have been altered as a result of a neurological dysfunction: the problem is one of learning *disability* rather than learning *incapacity*.

Although the participants in the seminar on brain-damaged children urged the adoption of an educationally relevant term, ". . . the term 'learning disability' did not meet the needs of the key professionals who were working in the field" (Hallahan and Cruickshank, 1973, p. 71). Nevertheless, the authors state in the preface to their text published eight years later:

> In the histories of special education and of the psychology of disability no issue has so suddenly captured the interest of parents, educators, pediatricians, psychologists, and the representatives of many other disciplines as that generally termed "learning disabilities." Poor as the term is, still illusive of any meaningful definition, it has caught

the imagination of those who serve children and of those who want children served. In the short period of a decade the term has attracted not only the attention of thousands of professional people and tens of thousands of parents, but it has also developed a mystique of its own. In many circles it is no longer the issue of learning disabilities, but the profession of Learning Disabilities. (1973, p. IX)

The term "learning disability" is also used by Johnson and Myklebust (1967). However, they qualify it by the inclusion of the adjective "psychoneurological" which was derived from the writings of Benton (1959) and Luria (1961). "Psychoneurological learning disability" means

. . . that the behavior has been disturbed as a result of a dysfunction in the brain and that the problem is one of altered processes, not of a generalized incapacity to learn. (1967, p. 8)

The waning of support for the use of the etiological term "minimal brain damage" has meant an increasing usage of the term "learning disabled" to categorize children exhibiting problems of a perceptual nature. The extent of this shift is well documented (Hallahan and Cruickshank, 1973, p. 136–58). It is evidenced by the proportion of published articles dealing with "brain damaged" children compared to those dealing with "learning disabled" children. In 1954, when the orientation toward a learning disorder first appeared, only .05 percent of the articles fell into this category; the remainder fell under the category of "brain damaged." By 1970, 67 percent of the published articles dealt with learning disabilities and 40 percent with brain damage. A breakdown of the articles according to sensory modality reveals, interestingly enough, that for the period 1961 through 1970, 44 percent of the articles on learning disabilities were concerned with visual and visual/motor behavior, while those concerned with auditory perceptual problems accounted for only half that figure.

Almost every definition of learning disability identifies the problem as arising from a dysfunction of the central nervous system. Children with problems arising from sensory deficits, environmental factors, poor teaching and learning situations, and mental retardation have been almost universally excluded from the category of the learning disabled. A typical definition of learning disability is the one offered by the Learning Disabilities Division Formulation Meeting of the National Council on Exceptional Children 1967:

A child with learning disabilities is one with adequate mental abilities, sensory processes and emotional stability who has a limited number of specific deficits in perceptive, integrative or expressive processes which severely impair learning efficiency. This includes children who have central nervous system dysfunction which is expressed primarily in impaired learning efficiency. (1967.)

The National Advisory Committee on Handicapped Children again clarifies the distinction:

Children with special learning disabilities exhibit a disorder in one or more of the basic psychological processes involved in understanding or in using spoken or written languages. These may be manifest in disorders of listening, thinking, talking, reading, writing, spelling or arithmetic. They include conditions which have been referred to as perceptual handicaps, brain injury, minimal brain dysfunction, dyslexia, developmental aphasia, etc. They do not include learning problems which are due primarily to visual, hearing, or motor handicaps, to mental retardation or to environmental disadvantage. (1968)

Although society has in the past excluded the environmentally disadvantaged child from special remedial measures for learning disabilities, a protest has been registered by Hallahan and Cruickshank. They argue that lack of experiences may cause a child to fail to learn or develop appropriate perceptual motor behaviors. They conclude:

It is, thus, our opinion that children who have learning problems due to environmental conditions should *not* be excluded from learning disability programs. (1973, p. 13)

Johnson and Myklebust have voiced the same sentiment:

A disadvantaged child, a child deprived of opportunity, will be deficient in various kinds of learning, despite even excellent potentialities. Hence when appraising deficiencies it is essential that opportunity be considered and evaluated. (1967, p. 7)

The term "learning disabilities" remains very much in use today despite the fact that its broad, noncategorical nature does little to define the nature of the problem. Since many otherwise "normal" children exhibit learning difficulties for a multitude of reasons, the term has tended to provide a "solution" for teachers and administrators unable to accommodate a particular child within the normal academic structure. As a result, children in classes for the learning disabled often constitute an extremely heterogeneous group with little in common except their difficulty in learning. This extreme heterogeneity precludes the effectiveness of a specialized intervention program. As Hallahan and Cruickshank state:

> Classes for mentally retarded children have often been criticized as being "dumping grounds" for the benefit of school personnel who could not solve certain problems in children. This criticism could apply a hundredfold to the typical public school program for learning disabilities. (1973, p. 8)

Identifying the Learning Disabled Child

Diagnostic evaluation of the specific learning disability is all important. Both Johnson and Myklebust (1967) and Hallahan and Cruickshank (1973) emphasize that the problems of the learning disabled child are fundamentally based on neurological function or dysfunction. The problem is to demonstrate that such dysfunction exists, in order to separate such children from others whose difficulties arise from sensory defects, mental retardation, or environmental factors.

The focus of our evaluation must be on difficulty in learning. Although perceptual processing problems and their resultant learning difficulties exist in children with mental retardation, cerebral palsy, or psychological disturbances, *the dominant factor in each of these types of problems is something other than the learning disability* (Johnson and Myklebust, 1967, p. 9). Children with perceptual processing problems share essentially normal functions in all areas of development *except learning skills.*

Our specific interest is in disorders involving auditory learning function. However, since language is the common denominator of all communicative learning, it is impossible to disregard the relationship of auditory to visual perception in the learning process. This is particularly important when one deals with children who experience difficulty in learning to read through an auditory approach. Most verbal learning involves the auditory system either directly through listening or indirectly through evoking auditory symbols by visual patterns. At some level, learning to read involves the establishment of equivalency between the two forms of information. This fact is evidenced by the very considerable problems deaf children have in learning to read. In subsequent chapters we shall examine the nature of auditory learning disorders which take the forms of speech articulation difficulty, receptive and expressive language processing problems, and reading difficulties.

REFERENCES

AYERS, A. J., 1975. "Sensorimotor foundations of academic ability," in *Perceptual and Learning Disabilities in Children*, Vol. 2 Research and Theory, eds. W. M. Cruickshank and D. P. Hallahan. Syracuse, N.Y.: Syracuse University Press.

BARR, D. R., 1972. *Auditory Perceptual Disorders, An Introduction.* Springfield, Ill.: Charles C Thomas.

BARSCH, R. H., 1967. "Achieving Perceptual Motor Efficiency." *Seattle Special Child Publications*, Chapter 14.

BENTON, A., 1959. *Right-left Discrimination and Finger Localization.* New York: Paul B. Hoeber.

BERRY, M. F., 1969. *Language Disorders of Children, The Basis and Diagnosis.* New York: Appleton-Century-Crofts.

BIRCH, H. (ed.), 1964. *Brain Damage in Children, The Biological and Social Aspects.* Baltimore: Williams and Wilkins.

CHELFANT, J. C. AND M. A. SCHEFFELIN, 1960. *Central Processing Functions in Children, A Review of Research*, Monograph No. 9., Bethesda, Md.: National Institute of Health, U.S. Dept. of Health, Education and Welfare.

CRUICKSHANK, W. M. (ed.), 1966. *The Teacher of Brain Injured Children.* Syracuse: Syracuse University Press.

FRY, D. B. AND E. WHETNALL, 1954. The auditory approach in the training of deaf children. *Lancet*, 1, 106.

HALLAHAN, D. P. AND W. M. CRUICKSHANK, 1973. *Psycho-educational Foundations of Learning Disabilities.* Englewood Cliffs, N.J.: Prentice-Hall.

HORTON, K. B., 1974. Infant Intervention and Language Learning. In R. L. Schiefelbusch and L. L. Lloyd (eds.) *Language Perspectives—Acquisition, Retardation and Intervention*, Chapter 18, pp. 469–491. Baltimore, Md.: University Park Press.

JOHNSON, D. J. AND H. R. MYKLEBUST, 1967. *Learning Disabilities, Educational Principles and Practices.* New York: Grune and Stratton.

KATZ, J. AND R. ILLMER, 1972. "Auditory Perception in Children with Learning Disabilities," in *Handbook of Clinical Audiology*, ed. J. Katz. Baltimore, Md.: Williams and Wilkins.

KIRK, S. A., April, 1963. "Behavioral Diagnosis and Remediation of Learning Disabilities," in *Proceedings of the Conference on Exploration into the Problems of the Perceptually Handicapped Child.* Chicago: First Annual Meeting, Vol. 1.

LENNEBERG, E. H., 1967. *Biological Foundations of Language.* New York: Wiley.

LLOYD, L. L., 1976. Discussants comment: Language and Communication Aspects in T. D. Tjossem (ed.) Intervention Strategies for High Risk Infants and Young Children. Baltimore, Md.: University Park Press.

LURIA, A., 1961. *The Role of Speech in the Regulation of Normal and Abnormal Behavior.* New York: Liveright Publishing Corp.

MYKLEBUST, H. R., 1964. *The Psychology of Deafness: Sensory Deprivation, Learning and Adjustment.* New York: Grune and Stratton.

NATIONAL ADVISORY COMMITTEE ON HANDICAPPED CHILDREN, January, 1968. First Annual Report, "Special Education for Handicapped Children. National Advisory Neurological Diseases and Stroke Council. *Human Communication and its Disorders—An Overview.* A report prepared and published by the Subcommittee on Human Communication and Its Disorders. National Institute of Health, HEW Bethesda, Maryland, 1969.

NATIONAL COUNCIL ON EXCEPTIONAL CHILDREN, April 1967. Learning Disabilities Division Formulation Meeting (CEC), St. Louis, Mo.

OAKLAND, T. AND F. C. WILLIAMS, 1971. *Auditory Perception Diagnosis and Development for Language and Reading Abilities.* Seattle, Wash.: Special Child Publications.

REAGAN, C. L., 1973. *Handbook of Auditory Perceptual Training.* Springfield, Ill.: Charles C Thomas.

WEDENBERG, E., 1961. Hearing measurements of infants. *Nord. Psychiat. Ridsker,* **191**, (2), 106.

9

Auditory Learning Difficulties as a Language Processing Function

We have seen that the efferent motor pathways operate to provide a control over the function of neural centers processing the incoming stream of information. In this way the efficiency of the system can be enhanced through the gating of neural impulses. Anticipated patterns of information can be processed with greater ease, since those patterns with the highest probability of occurrence according to the language rules are favored. The tuning of the system increases in effectiveness as the developing pattern progressively narrows down the probable future components of the whole. We stressed that this is an ongoing process tied to the stream of information conveyed to the system by the acoustic signal.

The Role of the Acoustic Speech Wave

It is obvious that the only link between speaker and listener is the acoustic speech wave. We know that this wave is created by a speaker who follows the rules of a particular language system. Whether these rules are directly encoded into the acoustic wave or whether they are mediated through the articulatory-resonant code remains to be conclusively demonstrated. Whichever may be the case, the fact remains that the speech wave pattern is symptomatic of the linguistic rules governing its production. It is the pattern of the acoustic wave which provides the listener with the cues he needs to identify the linguistic unit. As we saw in Chapter 2, for each unit of speech several cues are simultaneously present. These include not only segmental, but also suprasegmental cues. *In processing speech the emphasis is not on absolute judgments concerning the parameters of the signal, but on the changing relationships between them.*

> In some cases it is not the judgment of a single feature—like formant frequency—that has to be compared with a system, but rather a relation between succeeding parts of the utterance that has to be compared with, and fitted into, a system of such relationships. (Fry and Denes, 1959, p. 378)

The capacity of the system to make fine differentiation between the parameters of the acoustic signal has been shown to be present even in very young infants (see Chapter 7). However, the ability to identify meaningful patterns of information is dependent upon the restructuring of the segmental information from the directions encoded into the speech wave.

The decoding of the early instructions in the signal permits the internal synthesis of the pattern. As the contribution of the linguistic constraints increases, the dependency upon the acoustic signal progressively decreases. Thus, as unit length grows, several things happen: The statistical constraints become more strict, the possible choices of what the message might be decrease, and the accuracy of prediction increases. Each successive decoding permits the processing of increasingly larger segments. Since the acoustic wave carries information transmitted in parallel, the structure of the message is available at several levels at any given instant in time. As Fry and Denes explain,

> The segmental probabilities at any one linguistic level are determined not only by the preceding sequence at that level, but also by the preceding sequences of larger units. (1959, p. 380)

One can conceive of the system shifting down in gear as it moves away from strictly acoustic processing. This depends upon an increased momentum of language structure to carry it toward comprehension.

Thus, the initial task of the auditory perceptual system is the internalization of the speech wave pattern. We have seen that this necessitates the storage of undifferentiated information in what has been termed pre-categorical form.

To this point in the process, linguistic influences have not played a major role. However, all subsequent processing must take place in the light of language restructuring. The goal is not to analyze the speech, but to reconstruct the language pattern which serves to unlock the meaning attributed to the message. The critical requirement of the transfer of information from short- to long-term storage is the ability to segment it. This, we have seen, is dependent upon familiarity with, and utilization of, the linguistic structure. A listener is able to identify acceptable patterns of segmented information in the unsegmented acoustic stream because he is able to use the linguistic statistical probabilities to predict sequences which would be compatible with those already analyzed. Therefore, *auditory perception of speech appears to be first and foremost a language-processing function.* It does not seem to be the relatively independent function that some writers claim it to be. The process involves more than a combination of discrete auditory skills feeding information to a language system for linguistic analysis and interpretation. As Liberman categorically states, "speech is truly an integral part of language" (1970, p. 304).

Some insight into the nature of the relationship of the auditory stimulus to the derivation of meaning is provided by Aten:

> Language allows us to code information by categorizing stimuli. As we develop language normally, we form better codes which makes possible more elaborate categories for classifying and retaining what otherwise would be amorphous or meaningless discrete sensations. (1972, p. 116)

Blesser (1972, 1974) has shed some light on auditory discrimination of speech in relation to spoken language comprehension in adults. He performed a study involving pairs of male subjects who knew each other well. The subjects were required to learn to communicate with each other through an audiosystem which reversed the formant information arrangement of the speech signal. Thus, the F1 information pattern exchanged places with the F3 information pattern. The F2 transition was therefore reversed in direction of movement. The result of this formant inversion was to produce unintelligible speech. The destruction of intelligibility resulted from the distortion of the spectral information. Non-spectral, prosodic cues of pitch, stress, phonemic duration, and temporal order remained unchanged. The experiment studied both the capacity of the subjects to learn to understand the distorted speech and the learning strategies themselves. Change in discrimination performance was measured for phonemes, poly-syllabic words, and sentences after each of twenty learning trials.

Blesser's results are important to our discussion because he showed that in this experiment:

> Discrimination scores, phoneme identification scores, and single word identification scores were not correlated with one another or with the ability to converse. In other words, the subjects who showed marked ability to discriminate phonemes and could even get place uses right occasionally did not necessarily recognize words or sentences very well nor do well on conversation. The reverse was also true. (1974, pp. 141–42)

The results of the study further support the contention that the relationship between the ability to discriminate small units of speech and the ability to comprehend is at best tenuous. It would seem that when the language system is sophisticated and intact, we process information in large linguistic chunks. We use the envelope pattern of the acoustic event rather than the detailed constituents necessary for the processing of smaller components such as words, syllables, and phonemes. Yet, approaches to remediation of speech problems concentrate heavily on training at the phonetic level.

Disorders of the Perception of Speech

Such concepts as we have discussed present rather considerable problems. They do not fit well with the literature pertaining to the nature of auditory processing problems in children. With few notable exceptions, authors have attributed the *cause* of these problems to a dysfunction of one or more of several "auditory skills." This conclusion is easy to reach when difficulty on a specific task involving sequencing ability, auditory memory, or figure-ground separation can be demonstrated in a learning disabled child. The fact that speech is a continuous, sequenced event, that patterning arises from temporal order, and that short- and long-term memory are involved in the perception of speech is unassailable. *The problem does not arise, therefore, from dispute about the component aspects of processing, but centers on the question of whether a demonstrated dysfunction in any one or more of these abilities is causative or symptomatic in nature.*

In what may prove to be a classic article, Rees (1973) has reviewed the evidence which allegedly supports the assumption that discrete auditory processing skills are fundamental to the acquisition of language and academic learning. She stresses that speech perception results from the interaction of, rather than the summation of, the acoustical properties of the signal and the characteristics of the auditory perceptual system. Furthermore, Rees points out that it is not possible to predict speech perception from an analysis of the acoustic speech wave. We are not dealing with a simple, direct relationship between an acoustic event and a subsequent perception, but rather with the linguistic processing of the acoustic information.

We have considered in some detail the case presented by Liberman and his coworkers (1967, 1972) supporting the hypothesis that speech is processed in a speech mode according to special rules. We reviewed the evidence for categorical perception of the unsegmented acoustic stimulus, for the absence of phonemic units within the acoustic stream, for the probable relationship between the production of a speech sound and its subsequent perception, and for the specialization of the dominant hemisphere for the perception of rhythm and speech. We also noted that the rate at which we are able to process speech far exceeds the temporal resolving power of the ear for short duration sounds. Such evidence clearly argues against a process of speech perception which involves the sequential segmentation of the speech wave into phonetic or larger-sized units.

In Rees' opinion, the claim that a child must have a set of discrete auditory skills in order to learn to speak or read is at best weak. She believes that there exists "no hard core of evidence for any auditory factor underlying language or learning disorders" (p. 312). She states:

> The inescapable conclusion from that review is that the search for a single auditory skill, or even a set of auditory abilities, that is essential to language learning or impaired in all or most language-disordered children seems futile. We must, therefore, question the diagnostic value of tests that purport to isolate these skills, as well as the therapeutic value of clinical procedures designed to improve them. (1973, p. 312)

Similar arguments, specifically related to the nature of the difficulties encountered by some children in developing normal articulation patterns, were put forward earlier by Locke. Studies of acoustic signal processing in infants indicated that categorical discrimination between [ba] and [ga] can be definitely discriminated by three-month-old infants. Locke concludes:

> By way of summary, most of the work seems to suggest that acoustic signal processing is not an especially difficult task for infants and young children and that, to some degree, it follows the same laws and

respects and acknowledges the same phonemic features which adult speakers consider critical. (1971, p. 29)

With this in mind, it is ironical, Locke suggests, that we should subsequently consider efficient phoneme perception to be so difficult as to cause speech production problems in many children.

Powers (1971) made an extensive review of the literature pertaining to the symptomatology and etiology of functional (nonorganic) disorders of articulation. She also is of the opinion that there is no clear evidence to suggest that children with functional articulation problems perform at an inferior level on standard tests of speech discrimination and auditory memory. She does, however, cite research suggesting that some children may exhibit discrimination difficulties with sounds they misarticulate. Locke (1971) reports on a study of this particular phenomenon by Goldstein and Locke (1970), who assessed the phonemic production of five-year-old children using a picture-naming activity. Each child who misarticulated the name of a picture object was then asked six questions about the object. Three of the questions involved the correct articulation of the object's name, and three involved a mimicking of the child's incorrect articulation.

Locke stated:

> The results show an extremely interesting finding which I believe has value both for clinical application and for understanding children's phonology. When children were asked if the picture was the correctly articulated label, they answered "yes" with equal frequency regardless of whether they produced it as the examiner had. But when asked if the picture was the incorrect form of the label, those who said it correctly said "yes" 34 percent of the time; those who misarticulated the item said "yes" 67 percent of the time. (1971, p. 30)

This finding indicated that the child's perception of the correct model was independent of whether he himself correctly produced the sound. However, the children who misarticulated a phoneme usually perceived that misarticulation as correct when mimicked by others. This seems to indicate that a child may make an auditory discrimination which indicates perceived difference between sounds at an acoustic level, yet at the same time allows for the classification of the two sounds within the same phonemic category. Thus, two phonemes would be classified as allophones of a single phoneme. Locke and Goldstein (1971) and Warner and Locke (unpublished manuscript) in two subsequent studies, observed this behavior among children with deviant articulation patterns. Their results revealed what Locke (1971) considered to be

. . . the unmistakably clear finding that most of the time the phonemic contexts misarticulated and misperceived simply did not match. (1971, p. 30)

In other words, the children's problems in discrimination did not match their problems in articulation. Locke concludes that:

. . . the relationship between production and perception in misarticulating children is slight and somewhat obscure. (1971, p. 30)

Similar findings have been reported by McReynolds, Kohn, and Williams (1975), using the McDonald Deep Test of Articulation. They compared the performance of seven children with severe speech articulation difficulties with the performance of a control group of children with normal speech articulation. Their results indicated that the normal children performed well on both discrimination and articulatory production. As might be expected, the children with defective speech performed poorly on the production test; their scores on the discrimination test, however, approximated those of the normal children. Articulation defects did not prevent them from discriminating features and phonemes they themselves could not produce.

This apparent lack of correlation between the perception and the production of a sound might initially be interpreted as evidence in support of a nonmediated, passive form of perception. It certainly does not seem to support a motor theory of speech. We must bear in mind, however, that the task of discrimination requires a comparative judgment concerning the similarity or dissimilarity of the sound pattern received and the internal model. Such a judgment may be made at an acoustic or a linguistic level. A judgment by a child that two unlike phonemes are "the same" or "different" fails to indicate the criteria by which the decision is made. Thus, the decision may be purely acoustic or it may be a categorical, linguistic one. In the latter case, a judgment of "same" when two "different" phonemes are presented may not indicate that they are acoustically indiscriminable, but rather that the sounds have been classified by the child as allophonic variations of a single phoneme. The boundaries of the child's phonemic category may be very different from those of the adult. Monnin and Huntingdon (1974) showed, for example, that the children in their study who substituted /s/ for / ʃ / (sh) were not only unable to differentiate the phoneme boundaries, but failed to discriminate relevant differences between /sev/ and / ʃ ev/. They responded randomly, suggesting that for them /s/ and / ʃ / were allophones of a single phoneme and were thus included in the same category.

It is important, therefore, to remember, that for the children observed in the above studies, the lack of correspondence between perception and production is between a linguistic classification and an articulatory production, rather than between two distinctive articulatory patterns. The children with defective articulation accepted the acoustic pattern of both the correct and incorrect sound. They may have been making allophonic rather than phonemic categorization. This might result from the child's making an acoustic reference to his own defective articulation pattern. The process of normalization discussed earlier (p. 27) would account for a constant perception against a varying stimulus. The acceptance of both the correct and incorrect patterns, on the other hand, may have arisen from a mismatch between the criteria for judgment used by the tester and by the child. This would involve a difference in interpretation of the criteria by which stimuli were considered to be "correct" or "same"/"different."

Speech therapists frequently find that a child can differentiate easily between the correct and incorrect production of a speech sound in a test situation, yet fail to make such a differentiation in his own conversational speech. Frequently, neither more auditory discrimination training nor further articulation exercises resolves the problem of achieving carry-over. Such failure to make progress would be easier to understand if the difficulty lay at a language-coding level rather than at a more peripheral acoustic-processing level.

Some evidence of a possible relationship between articulation competency and linguistic coding has already appeared in the literature. Marquardt and Saxman (1972) studied the receptive processing of spoken language in children with functional articulatory disorders of speech. They investigated the relationship of language comprehension to auditory discrimination. Thirty speech-defective children with functional articulation problems were compared to a matched control group of children with normal speech. Children with misarticulated speech evidenced deficit performance on both the Carrow Auditory Test for Language Comprehension and the Wepman Auditory Discrimination Test. Their lower-than-normal performance on the two tests was found to be directly correlated with the number of articulatory errors they manifested.

An earlier study (Menyuk 1964) investigated the language production capabilities of children with nonorganic speech articulation problems. She demonstrated that children (between the ages of 3 years and 5 years 10 months) with deviant speech articulation also exhibited deviant use of the grammatical rules of spoken language. Shriner, Holloway, and Daniloff (1969) also examined this relationship. They tested the syntactic development of children with severe articulation problems. Thirty children, grades one through three, all with multiple articulation errors, were compared with a matched group of children with normal speech. The authors demonstrated that the group with defective speech articulation performed significantly worse in areas of grammatical completeness and used shorter sentences than the normal group. They state:

> The relationship reported between phonological errors and syntax deficits is, for age 3 through 13 years, 6 months, evidence against the relationship being a developmental process. Over a wide level of physical development, defective phonological production may induce and is associated with deficits in syntax. (1969, p. 322)

Evidence that the relationship between psychosensory function and language competency is not confined to the auditory system, but cuts across visual and intersensory (visual/auditory) systems, has been provided by McGrady and Olsen. They compared visual and auditory learning processes in normal and learning disabled children on a battery of thirteen psychosensory tests. The tests involved verbal and nonverbal stimuli. Their results indicated that the learning disabled group performed more poorly on tasks which utilized verbal stimuli regardless of the sensory modality used.

> Their problems manifest themselves to a greater extent in comprehension of language stimuli rather than perception of nonverbal stimuli. We were unable to discern that the parameters of sensory channel were useful in distinguishing between normal children and children with learning disabilities. (1970, p. 588)

They concluded that their evidence lent no support to the concept of learning difficulty as a perceptually based problem. McGrady and Olson state:

> The psychosensory test battery confirms that their primary learning disorders were verbal in nature; they had language disorders. (1970, p. 588)

If, as has been suggested, so-called auditory perceptual problems are intimately related to language-coding processes, then we may justifiably wonder where auditory processing fits into the picture.

Relating Component Aspects of Auditory Processing to Language Perception

Let us return to our consideration of the nature and role of the acoustic signal to see if we can gain further insight into the two apparently conflicting concepts of auditory learning problems.

Our view of auditory processing has stressed the role expectancy plays in a dynamic auditory system, constantly organizing and reorganizing itself to process information. The efficiency of the system is therefore heavily dependent upon the appropriateness of its readiness posture for a particular processing task. That readiness in turn is determined by the nature of the particular hypothesis generated by the system to cope with probable components of the speech signal. The process is heavily hierarchical allowing for correlations between several levels of language processing.

> The type of rules used to produce an internal response depends upon the situation. When the task is to shadow phonemes, the rules are probably operating at the level of distinctive features, whereas in ordinary speech perception, grammatical and semantical rules are in operation at the word and sentence level. That we can use different types of rules to guide this internal generation process seems intuitively obvious. (Weener, 1972, p. 157)

The linguistic hypotheses, therefore, influence the processing strategies adopted. These strategies must, however, have raw data upon which to operate. The raw data are characterized by the changing relationships of frequency and intensity over time. The record of these patterned changes must first be stored in short-term memory and then restructured into long-term memory.

Aspects of Auditory Processing

The acoustic information is transmitted as variations in the physical dimensions of frequency and intensity over time. The process of restructuring the input information through the several layers of linguistic complexity necessitates that the system first analyze the internal representation of the dimensions of the acoustic wave.

Adequate information processing requires that a person be able to discriminate among all those dimensions which have information-carrying value. Once the organism is able to make the necessary discriminations within a given dimension, that dimension can be used within higher-level organizational patterns to differentiate among larger units. (Weener, 1972, p. 159)

However, a person's ability to subject the data to such discriminatory assessment is in turn based upon his knowledge of the rules by which it can be segmented for comparative analysis. Thus, while our consideration of speech perception is strongly supportive of a holistic treatment of incoming speech data, it would be unrealistic to claim that there do not exist identifiable components in this process. The problem is that *at this time we do not know the specific nature of each function, nor do we understand how each interacts with other functions.* An attempt to describe the components makes apparent the fact that each is so intimately involved with all other aspects of processing that it is impossible to define any as truly autonomous. For example, the role played by memory in a sequencing task has been clearly demonstrated by Aaronson (1968). Immediately upon presenting the system with the task of processing the acoustic signal in the speech mode, each and every component is subservient to the language processing rules. Furthermore, since the system operates primarily in a parallel manner, processing of many single aspects occurs simultaneously. Let us now examine the major facets of the task of analyzing and resynthesizing the spoken message, but let us do so with caution, remembering the interrelationship of all aspects.

Awareness of acoustic stimuli. Awareness of the presence or absence of sound energy is the base level of processing. It provides the impetus for the auditory system to begin processing the stimulus. You may recall that in the acoustic theory proposed by Fant (1956), discussed in Chapter 5, one of the source characteristics of the speech wave is specified as silence; i.e., the awareness and recognition of absence of sound. Similarly, Utterly (Chapter 5, p. 112) defined the on/off phenomenon as one of the requirements for the principle of mathematical classification in the nervous system. The first question in analysis by a binary system would, therefore, involve the question "Is sound energy present?"

The role of awareness in stimulating the development of communicative abilities is fundamental, but it also has another equally important influence. This primitive function serves not only to permit the pre-linguistic and precognitive stages of language learning to occur (Blair, 1969), it also has significant psychological implications. Ramsdell (1970) stresses that awareness, or the "feeling of relationship with the world," is largely an auditory function. As I have explained elsewhere:

> It permits the ongoing monitoring of events providing an auditory background which satisfies our mind that we are participating in the changing world around us. It is a primitive monitoring process which serves as the coupler between the pattern of activity of our own system and that of the environment as a whole. (Sanders, 1975, p. 326)

In some children, this awareness may be impaired or may be slow to develop. Because of its relationship to the process of attention and to the establishment of emotional stability, and because of its role in pre-linguistic and precognitive stages of language development, this aspect of auditory processing plays a major part in speech and language processing. Until a child is aware that a sound has occurred or has changed, he will not be motivated to search for it. This has critical implications for congenitally peripherally hearing impaired children since for these infants the provision of early amplification is crucial to the learning of auditory/oral language.

Localization. Initially, search behavior is learned in terms of non-speech stimuli. It involves establishing correspondences between the auditory stimulus and the object, event, or person which gives rise to it. It requires intersensory integration of equivalent information so that eventually the total percept may be evoked by the auditory pattern alone. The learning of spoken language necessitates that a child be able to associate a particular pattern of acoustic stimuli, generated vocally by a speaker, first with concrete values (persons, things, or events) and later with abstract values (ideas). The ability to do this grows out of the ability to localize the sound-generating source and to link the properties of the acoustic event to the properties of the object giving rise to the event (Sanders 1971, 1976).

Sound localization is an important factor in establishing and maintaining an appropriate figure-ground relationship. Getman (1969) concluded from his work on visual and auditory performance in children that a child cannot consistently identify something which he cannot correctly localize. Certainly the localization of a sound source facilitates the process of attending to spoken language.

Attention. A person cannot shut out sound from his auditory system in the way that closing his eyes shuts out visual stimuli. However, as was explained earlier, the normal system is protected from being overwhelmed through the mechanism of selective processing. In order for a person to perceive a given message, he must be able to follow its developing pattern over time against a background of ongoing activity in the same medium. Selective attention allows for the control of this figure-ground relationship. The maintenance of the desired figure-ground relationship involves both the selective process of focusing attention and the sustaining of that focus for as long as is necessary for the identification and evaluation of the stimulus. Attention must be focused and then held. This latter ability is generally referred to as "attention span." Oakland and Williams combine these two aspects in a concise definition of auditory attention:

> Auditory attention. The ability to direct and sustain attention to sounds. Attention includes the ability to select a relevant stimulus from a background of irrelevant stimuli and to continue to attend selectively to this stimulus for an appropriate length of time. (1971, p. 10)

Listening cannot take place in the absence of selective attention. As we have discussed, attention is believed to involve some form of filtering out of redundant information resulting in the sharp focus of the desired pattern. A sensory overload results when this filtering function of the sensory system does not operate effectively. This frequently occurs at low levels of capacity. Information not relevant to the particular topic at hand fails to be gated out, or the integration of impulses gets out of phase, resulting in a nonfocused percept. As Wepman explained:

> It seems clear that one of the deficits of a highly distractible child who is not learning language is a fundamental inability to pay consistent attention to the succession of stimuli which makes the learning of verbal-auditory meanings possible. . . . (1969, p. 132)

Children with such difficulties are faced with a multitude of unweighted sensory stimuli. They are not able to attend to the relevant information.

The ability to pay attention is a developmental one. Hagen and Hale (1973) have proposed that the development takes place in the form of increasingly active participation by the child in the attention process. It is their opinion that some cognitive problems arise from the development of inadequate "attentional strategies." This view is elaborated in a review of the literature pertaining to the development of selective attention (Hagen and Kail, 1975). They discuss research into auditory attention conducted by Macoby (1967, 1969). The children studied were asked to repeat one of two messages presented simultaneously to opposite ears by a man's voice and a woman's voice. It was found that between the ages of 5-14 years a marked improvement occurred in the children's ability to repeat the correct words. Similarly, the number of cues a child could attend to increased with age.

Lewis also has reviewed research findings pertaining to attention and perception in the developing infant and young child. The evidence he discusses indicates that when cortical dysfunction exists, even subclinically, it shows up as a modification of the normal habituation-dishabituation response behavior to test stimuli. He calls for the use of this response decrement phenomenon as a diagnostic tool. He states:

> Given that response decrement to redundant information and recovery to change do measure central nervous system function and dysfunction, we now have a most valuable research tool for the investigator of individual differences in the opening months and even days of life. In addition it is not an age specific test, at least in the opening two or three years of life. (1975, p. 159)

Lewis believes that measures of attention afford great clinical promise for the development of highly sensitive and reliable procedures for identifying cortical involvement and pathology. Equally important, he suggests that the effectiveness of medical and educational intervention programs might be monitored in this way.

Auditory attention to speech is an aspect of language processing. It becomes evident that unfamiliarity or difficulty in applying the appropriate language-strategies will obstruct a child's attention to the speech stimulus. Difficulty in imposing an organizational structure on the incoming stream of information hinders control of the appropriate figure-ground relationship. This in turn impedes holding attention, since, as Weener points out:

> . . . discriminations which do not serve as the basis for higher level integrative frameworks drop out and are no longer even attended to. (1972, p. 160)

Thus competing environmental stimuli often cause problems for children with learning disabilities because of difficulty in holding figure-ground relationships for any length of time. The behavioral symptoms may include apparent unawareness by the child that he has been spoken to, or failure to persevere with a listening task (i.e., a story or programmed teaching tape) for more than a few minutes. The child often exhibits unusual levels of distractibility in response to competing environmental noise.

I have indicated that environmental noise levels in normal classrooms are surprisingly high, particularly in the early grades. (Sanders, 1965) My findings were later confirmed by Nober (1973). Nober and Nober (1975) investigated the effects of such classroom noise on the auditory discrimination performance of normal and learning-disabled children. They compared test results obtained in quiet and in the presence of classroom noise. The authors did not test under actual classroom conditions but simulated the noise situation by feeding taped classroom noise into a small test room. It is questionable whether this can be equated with live classroom acoustic conditions which may be more or less favorable to learning than a recording made in the same room. Nevertheless, the study showed that on the Wepman auditory discrimination test (1958) the learning-disabled children (ages 9.2 to 11.6 years) performed inferiorly to the normals both in quiet and in noise. Though both groups performed more poorly in noise than in quiet, the size of the difference for both groups was comparable. The authors acknowledge that they do not have evidence to indicate how critical this decrement in auditory discrimination may be for the child who is known to have reduced auditory processing abilities. However, it can reasonably be assumed that any factor which exerts a detrimental effect upon auditory perception among normal subjects is highly suspect as a contributing factor to the auditory learning difficulties of the child who has already demonstrated reduced ability for auditory learning under favorable conditions. In an earlier study, Lasky and Tobin (1973) investigated the effect of auditory distractors upon the ability of first-grade children to perform such common classroom tasks as responding verbally and in writing to verbally and visually-presented questions. Tape-recorded noise distractors were fed into the classroom during testing. The results of the study indicated that no significant difference occurred between the normal and learning-disabled children on tasks performed in quiet or in the presence of white noise. When linguistic distractors were used, however, the performance of the learning-disabled children deteriorated, while that of the normal children did not. Once again, the findings indicate that we appear to process linguistic information differently from non-linguistic information.

A comprehensive review of the research into the effect of distraction on learning-disabled children has been made by Hallahan (1975). It is of interest to note that in his summary and conclusions he states that "there is a tendency for auditory distractors to be more distracting than visual ones for learning-disabled children." However, studies involving auditory distraction are few in number. Since attention is so necessary to the perceptual processing of speech, we urgently need studies concerning the relationship between attention and levels of language development, as well as the effect of various types of auditory distractors on the learning process.

From these discussions, we can conclude that when normal control of the focus is impaired, or when the span is inadequate, the perception of speech and the processing of both receptive and expressive forms of language will be affected. Children with language-learning difficulties are frequently weak in the area of selective attention. Ironically, the deficit in language is at least a major contributing factor to this inattention. We have seen that the ability to anticipate appropriate language forms in the complex wave reduces the necessity of close attention to it. The inability to focus upon a desired stimulus leads to inconsistency of response to sound and difficulty in sustained listening. The problem is often quite accurately termed inattention, but unfortunately the child labeled "inattentive" may be thought to be deliberately so. Without an evaluation to determine the cause(s), he may be thought to lack the necessary intelligence to concentrate. "Inattentive" also applies to the difficult child who consistently daydreams or is easily distracted from the listening task by auditory or other sensory stimuli.

Differentiation Between Speech and Nonspeech. An early and critical decision to be made in processing the auditory stimulus is whether it shall be analyzed in a speech or nonspeech mode. (see Chapter 5) Some children with severe auditory processing difficulties may evidence disruption even at this early stage of processing. The inability to perform this task may originate from a figure-ground or an attention problem; or (in nonverbal children) it may arise from an inability to extract linguistic features from the stimulus.

The interrelationship between the various stages of processing is illustrated by the fact that, while speech perception is dependent upon the initial speech/nonspeech decision, that decision itself rests on the ability to process the signal as speech. Once the mode of processing has been chosen in favor of speech, the figure-ground relationship of speech vs. nonspeech sounds can be maintained. When we are required to listen to one speech sound versus another, difficulty in holding a desired figure-ground relationship is greatly increased.

Now the task of differentiating between auditory patterns, commonly referred to as auditory discrimination, can begin.

Auditory Discrimination. When we discriminate between auditory patterns, we do so on the basis of one or more perceptual variants. Sound patterns of various lengths and complexity differ initially in the acoustic information they yield, but even more importantly in their structural, linguistic nature. Auditory discrimination is, therefore, a major component of the process of resynthesizing the stimulus pattern. The unit of the pattern may be large, (the total acoustic event), or it may be small (a component of the total acoustic event). Differentiating between sound patterns depends initially upon the correct identification of the temporal relationships of both frequency and intensity components and the transitional cues arising from coarticulatory function. Like each of the other components discussed, discrimination is a complex, interdependent function. To differentiate between patterns necessitates awareness, focal attention, storage, and sequencing competency. The actual discrimination can occur only after the pattern has been internally resynthesized, when it is compared with the internal model and categorized.

Suprasegmental discrimination: Starke (1974) has hypothesized a close developmental relationship between the ability to discriminate differences in intonational contours of sentences and the ability to use the phonological rules of language for decoding. She refers to the contribution of several authors whose findings, she feels, lend support to the concept that suprasegmental information may develop first as cues to the emotive content of an utterance (Buhler, 1930). Subsequently, they provide an indication of stressed, high-information-bearing words (Macnamara, 1971; Brown, 1973), and at an even later developmental stage the child learns to use the suprasegmental information as cues to the syntactical structure of the language (Lieberman, 1967; McNeil, 1970). Such a developmental model is highly compatible with our earlier discussion (Chapter 5, p. 118 and Chapter 6, p. 139) of the probable role of the suprasegmental components in speech processing as authored by Martin (1972) in terms of early acoustic processing, and by Liberman (1970) and Liberman, Mattingly, and Turvey (1972) as it relates to grammatical restructuring. It lends support to the argument that many auditory perceptual problems may have their basis in language processing functions.

Segmental discriminations: Discrimination of the segmental components of speech, like that of the suprasegmental elements, calls for analysis of variations in patterns. It has been pointed out that such differential discriminations may be made at the syntactic level, at the morphological level, or even sometimes at the phonemic level. Most of the time processing shifts between levels, varying as a function of internal redundancy. The concept of auditory discrimination of speech should not be limited, therefore, to the differentiation and identification of phonemes. Berry, among others, has drawn attention to this and has questioned the value of testing discrimination for isolated sounds.

> Since auditory perception of language is posited on discrimination, not of single phonemes but of a series of phonemes in a prescribed order, we doubt the value of tests of recognition of isolated sounds. (1969, p. 123)

Speech is a continuum. Perception involves the ability to identify (discriminate) patterns within that continuum. The patterns programmed into it by the speaker adhere to definite linguistic rules. The auditory discrimination of meaningful speech must, therefore, be predicted by familiarity with those rules in operation. It must involve suprasegmental as well as segmental discriminations. This makes it imperative that the minimal perceptual unit be larger than the phoneme. (see p. 140)

Auditory discrimination also requires differentiation between syntactic and semantic patterns. The linguistic properties of words significantly affect their discriminability, a phenomenon directly related to the increased ability to apply a knowledge of language probabilities to the discrimination task. Maccoby (1967, 1969) showed that under test conditions correct identification of spoken messages increased over the age range of 5-14 years.

A number of researchers have investigated whether children can learn to repeat and increase their use of unfamiliar prepositional forms. Odom, Liebert, and Hill (1968) and Liebert, Odom, Hill, and Huff (1969) demonstrated that for children in the age groups 7-14 years, reinforcement significantly increased the use of English prepositional forms relative to a control group. However, no spontaneous novel forms (new rule) were evoked. This held true even when the subjects were asked to repeat the novel rewarded sentences immediately upon hearing them.

The effect of familiarity with appropriate language-processing strategies on the ability to discriminate syntactic constructions was further demonstrated in a subsequent study by Vasta and Liebert (1973). They showed that primary school children below the age of eight years were unable to discriminate unfamiliar syntactic constructions (article-noun-preposition). They studied sixteen first grade (6-7 years) and sixteen third grade (8-9 years) children. The children were exposed to rewarded new rule sentences interspersed with unrewarded sentences of English-rule or no-prepositional constructions. All subjects subsequently listened to sentences containing one of these three forms. They were asked to determine for each sentence whether it was the approved (would be rewarded) form. The results showed that third grade children were significantly superior to the first grade children in their ability to discriminate the "new rule" sentences. They also obtained superior scores in discriminating the other two constructions. The performance of the first grade children on the no prepositional sentences was below chance. This, Vasta and Liebert point out, suggests that an incorrect hypothesis (strategy) was being used to process these constructions. The authors suggest that the child appears to treat language learning as a problem-solving task involving the selection of a solution from among the alternatives he is able to generate.

Thus auditory discrimination of material which is processed linguistically is seen to be intimately linked with linguistic competence. Deficiency at the language coding level will be reflected in commensurate reduction of discrimination ability. Vasta and Liebert point out that while it has been shown that training in discrimination results in improved production using individual words (Mann and Baer, 1971), this appears not to hold true for larger syntactic structures. Thus Berry's (1969, p. 123) doubts concerning value of tests of isolated phoneme discrimination might well be extended to a careful assessment of the significance of what we learn from tests of auditory discrimination involving single words. The fact that the ability of Maccoby's (1967, 1969) subjects to discriminate the test words increased with age is probably a reflection of increasing language competency influencing the child's ability to attend in a selective manner. This relationship between attention and discrimination has been described by Zeaman and House (1973). They emphasize that the rate of acquisition of a discrimination ability is primarily determined by the child's ability to attend to the relevant features of the stimulus. If the stimulus is linguistic, the child will have to be familiar with the distinctive features of the pattern at some level of processing in order to be able to respond differentially.

Once again we see how interwoven the subcomponents of auditory processing really are.

Auditory Memory.   Because speech involves the spatiotemporal encoding of information, at no time do we have the total pattern before us. This holds true however small the unit may be. Even an isolated single speech sound has a beginning, a middle, and an end. *The production of a speech sound takes time.* The time factor necessitates that in receptive processing the internal representation of the acoustic event be held in storage as it is progressively resynthesized. Because the retention capacity of long-term memory appears to be limited to between five and nine units, "chunking" of components into larger, meaningful units is necessary. These phonetic chunks can then be restructured into other chunks of a higher linguistic order, turning patterns of phonemes into morphemes, patterns of morphemes into words, and words into sentence components. If a child has difficulty in learning the pattern of relationships which exist between component words, he will find it hard to identify meaning units. Deprived of the ability to "chunk" words into larger grammatical units, the child soon faces an overload situation, since he lacks the patterns which would further reduce dependency upon individual components. Instead of being able to process a limited number of logical auditory language units, the child has to attempt to retain a long string of single sounds or words, a task beyond the capacity of his short-term memory system. Some children may, for this reason, exhibit no difficulty in processing single words or two-word phrases, but may be unable to retain the earlier components of longer phrases for sufficient duration to be able to determine the whole pattern. By the time the final components have been processed, the earlier units have been lost, a problem which is accentuated with the size of the minimal meaning unit. It is not uncommon for children with such difficulties to react only to the latter components, giving answers which either evidence their failure to grasp the whole or simply reveal random guessing. Thus, a child with symptoms of auditory memory dysfunction may be unable to repeat orally or motorically an event involving more than two or three components. He may be unable to repeat a number of beats on a drum, or a series of numbers, single phonemes, or nonsense syllables. At a more advanced level, he may be unable to follow directions involving more than one component. The child may listen carefully to what is said, but when asked to reproduce the stimulus or to carry out the instructions, he can get no further than the last few components.

When we discussed memory, I referred to Crowder and Morton's (1969) concept of precategorical acoustic storage. The most recent few items of any sequence are held in a store which has a rapid decay factor of about two seconds. They envisaged that once the store is full, new items displace the stored ones as they arrive. In a later article, Crowder explains that:

> The only items that would be relatively free from degradation by subsequent input would be those which ordinarily are not followed by subsequent input—the last items in the list. The early list positions benefit from only very transient storage in PAS since new inputs are always following closely on their heels. (1972, p. 255)

Something of this nature is evidenced by children who on certain tasks are unable to restructure short-term memory information into the more highly coded long-term memory. They appear either unable to hold the sound image in PAS or unable to restructure it. This is apparent from the inability of the children to echo or mimic the complete pattern which they have just received.

Mimicry is sometimes referred to as "reauditorization." (Heasley, 1974, p. 31) It involves the rehearsal of the stimulus to extend the time it can be retained by permitting us to hear our own internal recording of the speech event. Conrad (1971), and Locke and Fehr (1972), have indicated that the recording of verbal stimuli into phonetic form is an almost inevitable occurrence in normal speakers over five years of age. The use of reauditorization or rehearsal serves, therefore, to overcome the limitations which PAS places on short-term memory. The question of whether the rehearsal process is performed using an iconic image of the acoustic phonetic event, or by use of motor-phonetic imagery, is an issue subject to the same types of arguments as the active vs. passive theories of speech perception.

Locke and Kutz (1975) have provided evidence which suggests that problems of speech articulation in some instances may *cause* a reduction in recall from short-term memory. This would further aggravate the problems of a child whose internal motor phonetic imagery is deviant from that for normal articulation. The authors argue that since sub-vocal rehearsal facilitates recall, any impediment to this ability reduces the amount and/or the accuracy of recall since the input will be inadequately or incorrectly rehearsed. In their comparative study of fifteen kindergarten children with correct articulation and fifteen with /w/ for /r/ substitution, Locke and Kutz showed that while all the children correctly perceived /w/ and /r/ when produced by the tester, those with deviant articulation of /r/ perceived their own tape-recorded misarticulation of /w/ for /r/ as /w/. On a memory task they made significantly more /w/-/r/ confusions in unspoken recall.

The authors point out that in considering the memory capacity of children with articulation disorders it is important to differentiate between memory for sensory form, which may be quite unimpaired, and memory for linguistic information. Since the latter involves recoding from the acoustic to the linguistic level, reference to a disordered code can result in *the efficient retention and recall of incorrect information.*

Locke and Kutz go on to suggest a more important question than whether children with disordered articulation exhibit recall problems. This consideration is where the missed items occur in the stimulus list. Since early items cannot be retained beyond the 2-3 sec. capacity of PAS (see Chapter 6, p. 150), their retention must be dependent upon restructuring into a linguistic code. The authors suggest, therefore, that problems arising from memory for speech will be manifest by errors in the latter third of the list. If, however, the errors occur primarily in the items which occur early in the list, it is likely that speech may not have been used effectively in recoding for memory. Studies by Fischler, Rundus, and Atkinson (1970) and by Meunier, Stanners, and Meunier (1971) are cited in support of this hypothesis.

Sequencing. Some children can recall the components of a pattern, but are unable to remember the order (i.e., temporal and spatial relationships of those components). We have seen how important temporal ordering of pattern components is to the determination of meaning at all levels from the morpheme to the sentence and even between sentences. The order of the units comprises an important constraint in determining meaning. At the segmental level, sequencing difficulties may take the form of sound reversals or reordering (frirɪdʒəreɪtə, sɪrəgɛt, æmɪnl, baɪkɪsl) it may occur in compound words (manmail; belldoor; waggonstation), or in combinations of words (allgone icecream; who it is?). Although some of these patterns may be acceptable at early stages of language development, their persistence beyond these stages is indicative of language processing difficulty.

Sequencing problems may also be manifest in the child's difficulty in reproducing the rhythmic structure of nonverbal acoustic patterns. He may have difficulty in deciding how many pairs of taps he hears in a sequence of pairs, or he may be unable to repeat an accented pattern of taps. Earlier, we discussed the finding that speech sounds are processed in the dominant hemisphere and nonspeech sounds are processed on the nondominant side (Chapter 5, p. 109). However, the processing of rhythmical patterns of nonspeech sounds performed by the dominant hemisphere is an exception supported by some research evidence (Robinson and Solomon, 1974; Halperin, Nachshon, and Carmon, 1973). That both speech and rhythmical pattern information (speech and nonspeech) are processed in the same hemisphere lends much support to Martin's (1972) hierarchical theory of speech processing. Robinson and Solomon state:

> It is probable that the left hemisphere is dominant for both nonspeech rhythms and speech stimuli because it is better able to do the hierarchical processing they both require. (1974, p. 510)

It is not surprising, therefore, to find that children who demonstrate auditory sequencing problems frequently experience difficulty in reproducing the rhythmical structure of nonverbal sounds. It is reasonable to assume that this difficulty will also be reflected in the processing of the suprasegmental components of speech which require the same hierarchical treatment. Since Martin's theory delegates an important role to the rhythmical patterning of speech, disturbance of this function may be expected to be associated with difficulties in the processing of segmental components. If a child cannot focus his attention on the sounds in a sequence, the pattern becomes distorted so that the meaning may be confused or lost. Sequencing is obviously intimately related to the structural rules of language processing, for it is these which determine the probabilities of the order in which segments can occur within any level.

Auditory Synthesis.   Because the acoustic signal generates a continuous stream of information, it is necessary for the auditory system to restructure the data into segmental units. This analysis of the components of the speech pattern is followed by synthesis of the segments into larger, more meaningful units at a higher level until the semantic whole has finally been restructured. Synthesis involves the blending of the bits of information into the various sized chunks which permit the identification of the total pattern.

Lehiste (1972) has suggested that primary auditory processing involves phonetic analysis (see Chapter 6, p. 140). This suggestion is not incompatible with Liberman's (1973) model which allows for the "folding in" of phonemic information to increase the predictability of the evolving pattern. This "folding in," which involves the simultaneous (parallel) transmission of several segments of phonemic information, assumes the capacity to synthesize information into complete patterns both at the production and perception stages. At some level, analysis of the acoustic stream requires segmentation and synthesis, since speech is perceived categorically. Once again, it needs to be recognized that the missing information, whether it be morphological, syntactical, or semantic, is generated from a knowledge of the structural rules of the language. Children with language processing problems frequently evidence difficulty in synthesizing the component units of a word or sentence because of low linguistic redundancy.

Synthesis is an aspect of speech processing that is closely related to attention. Unless the figure-ground control, an inherent aspect of attention, is possible, irrelevant information fails to be filtered out; this prevents the synthesis of the relevant components of the pattern.

Conclusions

The components of the auditory processing of speech have been treated by most educators and clinicians as separate abilities or skills. There is little doubt that each of the above-mentioned components is involved in some manner at some level in processing spoken language. It is, however, almost impossible to consider any one aspect as a discrete skill, since each is intimately involved with one or more other aspects. Williamson and Alexander have emphasized this point strongly:

> . . . these functions represent a complexity of overlapping, interrelated skills. Not only are they inseparable, one from the other, they are also inseparable from the expressive side of language except in an artificial sense. (1970, p. 2)

The claim that an auditory perceptual disorder can be pinpointed to a deficiency in one or more of these processes is, therefore, very hard to accept. The evidence offered in support of this claim is both equivocal and lacking of a sound theoretical model. The auditory skills concept finds little support from current research in linguistics, psycholinguistics, and psychoacoustics, yet the concept serves as the basis for most current approaches commonly used in remedial work with children with auditory learning difficulties. It is doubtful that poor performance on tests which claim to evaluate isolated skills reveals anything about a child's problem with the perception of spoken language. It is not even certain that a direct relationship exists between the two performances. However, it is perhaps premature to rule out completely the so-called "skills," for they may prove to be of symptomatic value, aiding us in our attempt to establish syndromes of auditory behavior which are indicative of particular underlying causative elements.

It has been with this goal in mind that investigators have sought to determine the nature of the posited relationship between speech discrimination and speech articulation. McReynolds, Kohn, and Williams (1975) argue that the contexts of the search have been too broad to permit the relationship to be specified. Certain hypotheses have been made about the nature of the relationship. Monnin and Huntingdon (1974), for example, provide evidence to suggest that where speech-sound-discrimination difficulty exists, it is specific to those sounds the child does not correctly produce. They feel, as does Powers (1971), that speech perception is tied to speech production. However, Monnin and Huntingdon question in which direction this relationship operates. On the basis of the discrete discrimination difficulties of speech defective children, they conclude that perception is probably more closely related to articulation than to the acoustic stimulus and that neuromotor mediation is necessary for phoneme identification (see our discussion of the motor theory, pp. 116–124).

Of particular interest is the evidence that suggests a significant correlation between speech articulation difficulties and lowered language comprehension (Marquardt and Saxman, 1972). *The degree of language deficiency appears to be tied to the number of incorrectly articulated phonemes.* That the child with multiple speech articulation problems evidences language deviancy in both receptive and expressive stages (Shriner et al., 1969) emphasizes the intimate relationship of speech articulation to the other levels of perceptual processing both for receptive and expressive language.

It is not our intention to argue that all speech articulation disorders are rooted in language processing difficulties. Certainly, some articulatory errors arise from poor speech models, motor coordination problems, and so on. What we do strongly recommend is that those articulation disorders labelled "functional," particularly where multiple errors are involved, be examined within the context of the language system. The importance of this approach goes beyond its implication for remedial strategies in speech therapy. The uncovering of the actual nature of auditory perceptual problems will not only provide a key to the management of nonorganic disorders of speech articulation, but will also shed light on the problems encountered by so many children in learning to read.

REFERENCES

AARONSON, D., 1968. Temporal course of perception in an immediate recall task. *J. Exp. Psychol.*, **76,** 129–40.

ATEN, J. L., 1972. "Auditory Memory and Auditory Sequencing," in *The Proceedings of the First Annual Memphis State University Symposium on Auditory Processing and Learning Disabilities*, ed. D. L. Rampp. Memphis, Tenn.: State Univ. Press.

BERRY, M. F., 1969. *Language Disorders in Children.* New York: Appleton-Century-Crofts.

BLAIR, FRANCIS X., December, 1969. Programming for auditorily disabled children. *Exceptional Child*, **36,** 262.

BLESSER, B., 1974. "Discussion: Aiding Speech Reception of Hearing Impaired Listener," Chapter 6 in *Sensory Capabilities of Hearing Impaired Children*, ed. R. Stark. Baltimore: University Park Press.

BLESSER, B., 1972. Speech perception under conditions of spectral transformation: I Phonetic characteristics, *J. Sp. Hrg. Res.*, **15,** (1), 5–41.

BLESSER, B., 1969. "Perception of spectrally rotated speech," Unpublished doctoral dissertation, M.I.T.

BROADBENT, D. E., 1961. Attention and the perception of speech. *Scientific American*, 143–51.

BROWN, R., 1973. *A First Language: The Early Stages.* Cambridge, Mass.: Harvard Univ. Press.

BUHLER, C., 1930. *The First Year of Life.* New York: Day.

CONRAD, R., 1971. The chronology of the development of covert speech in children. *Develop. Psych.*, **5,** 398–405,

CROWDER, R. G., 1972. "Visual and Auditory Memory," in *Language by Ear and by Eye*, eds. J. F. Kavanagh and I. G. Mattingly. Cambridge, Mass.: The M.I.T Press.

FANT, G., 1960. *Acoustic Theory of Speech Production.* Gravenhage: Mouton.

FISHLER, I., D. RUNDUS, AND R. ATKINSON, 1970. Effects of overt rehearsal processes on face recall. *Psychonomic Science*, **19,** 249–50.

FRY, D. B. AND P. DENES, 1959. "An Analogue of the Speech Recognition Process," in *Mechanization of Thought Processes*, National Physical Laboratory Symposium, 1, (10). London: Her Majesty's Stationery Office.

GETMAN, G. N., August, 1969. Explorations into visual-auditory space. *Optometric Extension Program*, **12**, 41, 76.

GOLDSTEIN, J. I. AND J. L. LOCKE (Unpublished manuscript). "Children's Identification and Discrimination of Phonemes."

HAGEN, J. W. AND G. A. HALE, 1973. "The Development of Attention in Children," in *Minnesota Symposia on Child Development*, ed. A. Pick. **7**, Minneapolis: Univ. of Minnesota Press.

HAGEN, J. W. AND R. V. KAIL, JR., 1975. "The Role of Attention in Perceptual and Cognitive Development," Chapter 5 in *Perceptual and Learning Disabilities in Children*, **2**, eds. W. M. Cruickshank and D. P. Hallahan. Syracuse, N.Y.: Syracuse Univ. Press.

HALLAHAN, D. P., 1975. "Distractibility in the Learning-Disabled Child," in *Perceptual and Learning Disabilities in Children*, **2**, eds. W. M. Cruickshank and D. P. Hallahan. Syracuse, N.Y.: Syracuse Univ. Press.

HALPERIN, Y., I. NACHSHON, AND A. CARMON, 1973. Shift of ear superiority in dichotic listening to temporally patterned nonverbal stimuli. *J. Acoust. Soc. Amer.*, **53**, 46–50.

HEASLEY, B. E., 1974. *Auditory Perceptual Disorders and Remediation.* Springfield, Ill.: Charles C Thomas.

JOHNSTON, D. J. AND H. R. MYKLEBUST, 1967. *Learning Disabilities: Educational Principles and Practices.* New York: Grune and Stratton.

LASKY, E. Z. AND H. TOBIN, 1973. Linguistic and nonlinguistic competing message effects. *J. of Learn. Disabil.* **6**, 243–50.

LEWIS, M., 1975. "The development of attention and perception in the infant and young child," in *Perceptual and Learning Disabilities in Children*, **2**, Research and Theory, eds. W. M. Cruickshank and D. P. Hallahan. Syracuse, N. Y.: Syracuse Univ. Press. Chapter 3, 137–62.

LIBERMAN, A. M., 1970. The grammars of speech and language. *Cognitive Psych.*, **1**, 301–23.

LIBERMAN, A. M., F. S. COOPER, D. P. SHANKWEILER, AND M. STUDDERT-KENNEDY, 1967. Perception of the speech code. *Psych. Rev.*, **74**, 431–61.

LIBERMAN, A. M., I. G. MATTINGLY, AND M. T. TURVEY, 1972. "Language codes and memory codes," in *Coding Processes in Human Memory*, eds. A. W. Melton and E. Martin. Washington, D.C.: V. H. Winston.

LIEBERMAN, P., 1967. "Intonation and the syntactic processing of speech," in *Models for the Perception of Speech and Visual Form*, ed. W. Walthen-Dunn. Cambridge, Mass.: The M.I.T. Press.

LIEBERT, R. M., R. D. ODOM, J. H. HILL, AND R. L. HUFF, 1969. Effects of age and rule familiarity on the production of modeled language constructions. *Develop Psych.*, **1**, 108–12.

LOCKE, J. L., Spring 1971. The child's acquisition of phonetic behavior. *Acta Symbolica*, **2**, (1), 28–32.

LOCKE, J. L. AND F. FEHR, 1972. Subvocalization of heard or seen words prior to spoken or written recall. *Amer. J. of Psychol.*, **85**, 63–68.

LOCKE, J. L. AND J. I. GOLDSTEIN, 1971. Children's identification and discrimination of phonemes. *British J. of Disorders of Communication*, **6**, 107–12.

LOCKE, J. L. AND K. J. KUTZ, 1975. Memory for speech and speech for memory. *J. Sp. Hrg. Res.*, **18**, (1) 176–91.

MACCOBY, E. E., 1967. "Selective Auditory Attention in Children," in *Advances in Child Development and Behavior*, **3**, eds. L. P. Lippsitt and C. C. Spiker. New York: Academic Press.

MACCOBY, E. E., 1969. "The development of stimulus selection," in *Minnesota Symposia on Child Psychology*, **3**, ed. J. P. Hill. Minneapolis: Univ. of Minnesota Press.

MACNAMARA, J., 1972. Cognitive basis of language learning in infants. *Psych. Rev.*, **79**, 1–13.

MANN, R. AND D. M. BAER, 1971. The effects of receptive language training on articulation. *J. of Applied Behav. Anal.*, **4**, 291–98.

MARQUARDT, T. P. AND J. H. SAXMAN, 1972. Language comprehension and auditory discrimination in articulation deficient kindergarten children. *J. Sp. Hrg. Res.*, **15**, (2), 383–89.

MARTIN, J. G., 1972. Rhythmic hierarchical versus serial structure in speech and other behavior. *Psych. Rev.*, **79**, 487–509.

McGRADY, H. J. AND D. A. OLSEN, April, 1970. Visual and auditory learning processes in normal children and children with specific learning disabilities. *Exceptional Children*, 581–88.

McNEILL, D., 1970. The Acquisition of Language. New York: Harper & Row.

McREYNOLDS, L. V., J. KOHN, AND G. C. WILLIAMS, 1975. Articulatory defective children's discrimination of their production errors. *J. Speech and Hrg. Dis.*, **40**, (3), 327–38.

MENYUK, P., 1964. Comparison of grammar of children with functionally deviant and normal speech. *J. Sp. Hrg. Res.*, **7**, 109–21.

MEUNIER, G., R. STAUNERS, AND J. MEUNIER, 1969. Pronounceability, rehearsal time, and the primacy effect of free recall. *J. Exp. Psychol.*, **1**, 36–34.

MONNIN, L. M. AND D. A. HUNTINGDON, 1974. Relationship of articulatory defects to speech-sound identification. *J. Sp. Hrg. Res.*, **17**, (3), 352–66.

MORSE, P. A., 1974. "Infant Speech Perception: A preliminary model and review of the literature," in *Language Perspectives-Acquisition, Retardation, and Intervention*, eds. R. L. Schiefelbusch and L. L. Lloyd. Baltimore: University Park Press.

NOBER, L. W., 1973. Auditory discrimination and classroom noise. *Reading Teacher*, **27**, (3), 288–91.

NOBER, L. W. AND E. H. NOBER, December, 1975. Auditory discrimination of learning disabled children in quiet and classroom noise. *J. Learning Disabil.*, **8**, (10), 57–60.

OAKLAND, T. AND F. C. WILLIAMS, 1971. *Auditory Perception*. Seattle, Washington: Special Child Publications.

ODON, R. D., R. M. LIEBERT, AND J. H. HILL, 1968. The effects of modeling cues, reward and attentional set on the production of grammatical and ungrammatical syntactic constructions. *J. Exper. Psychol.*, **6**, 131–40.

POWERS, M. H., 1971. "Clinical and Educational Procedures in Functional Disorders of Articulation," in *Handbook of Speech Pathology and Audiology*, ed. L. E. Travis. New York: Appleton-Century-Crofts.

RAMSDELL, D. A., 1970. "The Psychology of the Hard-of-hearing and Deafened Adult," in *Hearing and Deafness*, 3rd ed., eds. H. Davis and S. R. Silverman. New York: Holt, Rinehart, and Winston.

REES, N. S., August, 1973. Auditory processing factors in language disorders: The view from Procrustes bed, *J. Sp. Hrg. Dis.*, **38**, (3), 304–15.

ROBINSON, G. M. AND D. J. SOLOMON, 1974. Rhythm is processed by the speech hemisphere. *J. Exp. Psychol.*, **102**, (3), 508–11.

SANDERS, D. A., 1971. *Aural Rehabilitation*. Englewood Cliffs, N.J.: Prentice-Hall.

SANDERS, D. A., 1975. "Hearing Aid Orientation and Counseling," in *Amplification for the Hearing Impaired*, ed. M. Pollack. New York: Grune and Stratten.

SANDERS, D. A., 1976. "A Model of Communication," in *Communication Strategies—Assessment and Intervention*, ed. L. L. Lloyd. Baltimore: University Park Press.

SANDERS, D. A., 1965. Noise conditions in normal school classrooms. *Exceptional Child*, **31**, 344–53.

SHRINER, T. H., M. S. HOLLOWAY, AND R. G. DANILOFF, 1969. The relationship between articulatory defects and syntax in speech defective children. *J. Sp. Hrg. Res.*, **12**, (2), 319–25.

STARK, R. W., Ed., 1974. "Looking to the Future: Overview and Preview," Chapter 11 in *Sensory Capabilities of Hearing Impaired Children*. Baltimore: University Park Press.

UTTLEY, A. M., 1957. "Conditional Probability Computing in a Nervous System," in *Mechanization of Thought Processes*. London: Her Majesty's Stationery Office.

VASTA, R. AND M. LIEBERT, 1973. Auditory discrimination of novel prepositional constructions as a function of age and syntactic background. *Devel. Psychol.*, **9**, 79–81.

WAGNER, J. C. AND J. L. LOCKE, (unpublished manuscript). Children's Phoneme Discrimination in Familiar and Unfamiliar Phonetic Contexts.

WEENER, P., 1972. "Toward a Developmental Model of Auditory Processes," in *The Proceedings of the First Annual Memphis University Symposium on Processing and Learning Disabilities*, ed. D. L. Rampp. Memphis, Tenn.: Memphis State Univ. Press.

WEPMAN, J. M., 1969. "Approaches to the Analysis of Aphasia," in *Human Communication and its Disorders: An Overview*. A report prepared and published by the Subcommittee on Human Communication and its Disorders. National Advisory Neurological Diseases and Stroke Council. National Institute of Health. Bethesda, Md.: HEW.

WEPMAN, J. M., 1958. *Auditory Discrimination Test*. Chicago: Language Research Associates.

WILLIAMSON, D. G. AND R. ALEXANDER, 1975. Central Auditory Abilities. *Maico Audiological Library Series*, **13**, (7).

10

Speech Perception and Reading

Speech perception and reading both involve the processing of verbal language transmitted in a coded form. Although the physical stimuli received by the auditory and visual systems are of a different nature, they both evoke the same linguistic percepts. At some level of processing there must be, therefore, an equivalence between the two intake modes.

When we considered theories of speech perception, we saw that the fundamental issue was whether speech could be perceived directly from the acoustic event or whether it was mediated by an active, referential process. This question has its parallel in reading theories. These must address themselves to the problem of whether we can comprehend the message directly from the printed letters, or whether the visual patterns must first be referred to (mediated by) the speech code. Current literature gives little support to the non-mediated, or passive, theories of processing the printed word. If we assume that the mediated theories are most viable, the further question is raised concerning whether that mediation is articulatory in nature or is achieved through evoking an auditory image of the sounds or letters being read.

Finally, the problems encountered by children learning to read raise exactly the same question we have just considered in relation to auditory processing problems: Are these difficulties causative or symptomatic? Does the difficulty arise at the level of visual coding or at the level of linguistic processing?

Passive or Active Decoding

Passive Theory.   In comparing speech perception and reading, we must note one important exception to our comparable processes—namely, that while the understanding of spoken language is acquired naturally by the child, reading has to be learned. We must also keep in mind that many apparently normal children have considerable difficulty in learning to read, and some fail completely. Addressing this disparity, Mattingly (1972) points out that speech is a primary linguistic activity, while reading is a secondary activity grafted onto the primary linguistic code. Reading is, therefore, heavily dependent upon the reader's familiarity with the primary linguistic code, usually acquired through auditory perception. Congenitally deaf children experience considerable difficulty in learning to read because of marked retardation in verbal language (Stark, 1974; Gibson, Shurcliff, and Yonas, 1970; Kavanaugh, 1968; Myklebust, 1970).

In the most literal sense, a passive explanation of the reading process holds that the semantic value of the written stimulus is identified directly from the printed stimulus (Kohlers, 1970). This would require the establishment of a template for every word in our lexicon and the generation of a new internal image for every new word encountered. This is possible, for as Gough (1972, p. 335) points out, the written form of many languages does not use an alphabet system. The reader of these languages must make a direct association between the written symbol and the meaning of each word. But this procedure is a highly inefficient one, since the reader already has a large vocabulary of words and a phonological system for restructuring them.

A more logical explanation of the reading process lies, therefore, in the identification of the meaning of the written word by way of the sound elements of spoken language.

> If auditory imagery is a genuine biological phenomenon, then the sounds of speech must be included in the definition, and silent reading can be accompanied by a succession of auditory speech images that might have the same physiological function in the reading process as does silent articulation. (Conrad, 1972, p. 207)

This correlation can be envisaged as a simple, direct process involving the learning of a set of symbols which are equivalent to the auditory counter-parts (spoken words and sounds) the child already uses effectively. Once the equivalence has been learned, the segmentation of the written message into higher-order linguistic units, and thus to meaning, can occur as an automatic, preconscious transposition.

> The process of learning to read is the process of transfer from the auditory signs for language signals which the child has already learned to the new visual signs for the same signals. (Fries, 1962, p. xv)

Gough has expanded on this passive theory in a manner compatible with both the acoustic and the neurological theories of speech perception considered earlier (Chapter 5, p. 106). He hypothesized that the reader converts the visual pattern information by mapping it onto what he calls "systematic phonemes."

> Systematic phonemes are abstract entities that are related to the sounds of the language—the phonetic segments—only by means of a complex system of phonological rules. Thus it is easy to imagine that formation of a string of systematic phonemes would necessarily take place at some temporal distance (i.e., some time before) the posting of motor commands. . . . (1972, p. 337)

Active Theory.   Active theories of reading draw upon the motor and analysis-by-synthesis models of speech processing.

The motor theory, as applied to reading, postulates that in processing the printed or written word we overtly or covertly articulate the speech sounds, producing either vocal or subvocal speech. Considerable research has been carried out to determine whether articulation occurs during silent reading (McGuigan, 1970; Edfeldt, 1960). This research has utilized the procedure of recording muscle potentials in the speech organs during reading (electromyography, EMG). The evidence clearly indicates that the speech musculature is almost always involved in the act of silent reading. This involvement can occur all the way down to the peripheral articulators. As anyone who has watched first-grade readers struggling with primary reading material will know, beginning readers find it impossible to read without saying the words, even though they may be restrained from actually vocalizing them. We observe, similarly, that when the material is difficult to read because the content is unfamiliar, because it is poorly expressed, is syntactically abstruse, or is not clearly legible, we experience a greater need to articulate than when the material is simple. This has been documented by several researchers (Hardyck and Petrinovitch, 1970; Novikova, 1966; Edfeldt, 1960).

When Liberman (1967) first proposed the motor theory of speech he postulated that the articulatory restructuring of the message from the acoustic stimulus occurred at the level of the articulators. The theory was subsequently revised to allow this restructuring to occur at the level of neuromotor commands to the speech organs, without reference to the periphery. Fant's model (Chapter 5, p. 104) also allows for both active and passive involvement of the speech organs in speech decoding. Stevens and Halle (1967) agree that reference to articulation for speech production generally is handled at a pre-motor level. They suggest that inhibition of the motor pathway occurs, resulting in the production of an auditory-hypothetical pattern rather than a motor-articulatory pattern. The suppression of motor activity and the use of an auditory-hypothetical pattern during silent reading is quite conceivable.

In summarizing the research dealing with the role of articulation in silent reading, Conrad (1972) concludes that the adult reader almost always recodes the visual input stimulus into its phonological correlates. He even does this under experimental conditions where neither memorization nor comprehension is required, for example in a pure visual-encoding task using verbal stimuli (e.g., searching printed material for each occurrence of a given letter), in which transposition of the visual information into auditory coding is detrimental to performance (Corcoran and Rouse, 1970, Corcoran, 1971). However, Conrad stresses that this phonological recoding of the printed word constitutes a preference, not a necessity (1972, p. 224). He concludes:

> In the end then, our view is that reading is most certainly possible with no phonology involved at all, but with phonology it is a great deal easier. Our written language is a system for describing and distinguishing the sounds of spoken language. Informationally printed words are far more economic than figurative descriptions. Perceptually they are far more discriminable as speech sounds than as pictures. In our view this begins to add up to reasonable grounds for all of that neurological bother involved in transducing the little lines on paper into the language of that inward ear which is the next door neighbor to the "inward eye." (1972, p. 237)

Reading as Language Processing

Despite the similarities, Mattingly (1972) refuses to accept the explanation of the reading process as a simple procedure parallel to listening. He instead describes the reading process as a language-based skill. He views the primary activities of speaking and listening as processes involving the generation of an internal linguistic pattern. The semantic and phonetic representations in the pattern are linked by the grammatical rules of the language. This represents an application of the principles proposed by Liberman (1970) and Liberman, Mattingly, and Turvey (1972) discussed on pages 116–24. The production of a sentence thus requires the synthesis of a phonetic pattern linked to meaning by syntactic rules. Similarly, the comprehension of a sentence is dependent, in Mattingly's view, upon the ability of the receiver to generate a sentence compatible with the decoded phonetic pattern. Continuing with this analysis-by-synthesis approach, Mattingly then reminds us that during the entire process the listener remains aware of the meaning of the utterance, how it sounds, and any other linguistic characteristics it may have. This awareness (which Mattingly believes allows the listener, when necessary, to consciously control linguistic activity) also permits the speaker/listener to reflect upon his linguistic experience.

> . . . by virtue of this awareness he has an internal image of the utterance and this image probably owes more to the phonological level of representation than to any other level. (1972, p. 140)

In Mattingly's opinion reading is a language skill not analogous to listening. He conceives of the process as an active, generative one in which knowledge of the language structure is used to synthesize the sentence according to the desire of the writer. Instead of the written text's serving as a clue to articulatory gestures, Mattingly believes that it consists of "discrete units clearly separated from the text" (p. 141). He believes that the correspondence between these meaning units and the phonology is a simple, direct one. The length of the segment processed will vary with the redundancy of the material just as is believed to be the case in speech processing. For the most part, Mattingly believes, we depend upon the recognition of large patterns to process long strings of units as segments. This is not unlike a look-and-say approach, but the size of the patterns that can be sight-read by an experienced reader probably includes whole phrases whose images are familiar to that reader. However, when the reader encounters a new pattern, he is forced to identify the word by relating the new image to his spoken vocabulary in which, hopefully, we will find the meaning. If the meaning eludes him, he must develop a new phonological pattern for that word together with an appropriate semantic value. The process is described by Mattingly in this way:

The skilled reader, however, does not need complete phonological information and probably does not use all of the limited information available to him. The reason for this is that the preliminary phonological representation serves only to control the next step of the operation, the actual synthesis of the sentence. By means of the same primary linguistic competence he uses in speaking and listening, the reader endeavors to produce a sentence that will be consistent with its context and with this preliminary representation. In order to do this, he needs, not complete phonological information, but only enough to exclude all other sentences which would fit the context. As he synthesizes the sentence, the reader derives the appropriate semantic representation, and so understands what the writer is trying to say. (1972, p. 142)

In Mattingly's model the major emphasis is not on the ability to recover meaning from the written word by internally generating its auditory correlate. The reading process is seen as the ability to synthesize a sentence which satisfies upward and downward criteria; that is, the sentence generated must fit both the lower-order phonological constraint pattern and the higher-order syntactic and semantic constraints governed by context.

Learning to read is seen as a language processing task. According to this theory, it is neither the simple determination of auditory and visual pattern correspondences nor the recoding of visual patterns into articulatory patterns by way of phonological correlates. Reading, Mattingly suggests, is

. . . a deliberately acquired language-based skill, dependent upon the speaker-hearer's awareness of certain aspects of primary linguistic activity. By virtue of this linguistic awareness, written text initiates the synthetic linguistic process common to both reading and speech, enabling the reader to get the writer's message and so to recognize what has been written. (1972, p. 145)

Speech and Reading Processing Problems

There has been a shift away from explaining speech and reading in terms of the acquisition of a discrete set of auditory-visual/auditory skills. The emphasis now seems to be upon consideration of both listening and reading as language processes. Adequate performance in each of these skills appears to be directly related to the child's ability to acquire and use

appropriate linguistic strategies. Mattingly suggested that linguistic awareness, that is, the ability to internalize the stimulus information and then to
know and use the appropriate modes of processing (linguistic strategies),
underlies the ability to read. Rees (1973) concurs that the understanding
of spoken language is dependent upon a knowledge of linguistic structure.

Throughout the text we have seen evidence of how the structure and
function of the auditory perceptual system has evolved to adjust its neurophysiological postures to reflect certain expectancies about the signal pattern. Those expectancies are derived from linguistic rules operating equally
on speaker and listener in a given language culture. As a result of those
rules, the acoustic event contributes simultaneously, in a parallel manner,
information concerning several linguistic levels of the message. The evidence on segmentation leads to the conclusion that the system segments
the information into linguistic units of various lengths depending upon its
ability to predict the message. The particular rules we use to process the
signal pattern necessarily influence how we perceive it, and whether we
accept the hypothesis we generate at a specific level of processing.

With these observations in mind, it is reasonable to examine problems in the processing of speech or written material as problems in applying appropriate language strategies to the incoming stimulus pattern. If
the linguistic awareness factor that Mattingly proposes varies within the
population, as do all human attributes, *one must suppose the existence of
a threshold of awareness below which speech and reading problems begin
to show up.* (Mattingly 1972) Individual differences manifest by "an ear
for languages" or by unusual verbal ability might be similarly explained.
We might also hypothesize that, since the processes of speech perception
and reading are dominated by the generative language abilities of the
listener, subtle neural dysfunction or faulty early learning could conceiv-
ably account for many of the highly persistent problems in speech and in
reading.

During our discussion of speech articulation, we pointed out that,
since speech is encoded in a layered manner, a child with an articulation
problem may experience no difficulty in decoding at one level yet considerable difficulty at another. For example, a child who easily discriminates
between two sounds at a phonetic level may be unable to assign them different values at a phonemic level. Conversely, a child who can process
at the semantic level (he understands what is said) is not necessarily able
to perceive or articulate individual phonemes correctly. This was true
in the formant inversion study by Blesser (1974). The child's processing
competence will depend upon his level of linguistic awareness. Inability
to perceive phonemes is analogous to the holophrasic utterances of infants,
such as "wheresitgone?" "whasat?" "supestairs" (it is up the stairs), which

cannot be further segmented because of lack of linguistic awareness. My own lack of linguistic awareness while living in Norway led me to believe that the words "instead of" were a single word in Norwegian, "istedenfor." Not until I began to write the language did I become linguistically aware that it is comprised of three words "i steden for."

When we discussed segmentation (Chapter 6, p. 140), we were led by Lehiste (1972) to conclude that although perception of phoneme-like units is possible for adults, the syllable is probably the most likely candidate for the minimal segment size in language processing. Nevertheless, the ability to segment syllables into phonemes plays an important role in the phonic approach to reading and an even more important role in the predominant techniques of speech correction for articulation disorders. Yet as Levin (1972, p. 322) points out, the inability to analyze syllables into segments is one of the characteristics of nonreaders at first- and second-grade levels. He emphasizes that phonemes can probably be identified only by analyzing syllables already perceived.

If this is so, and the evidence seems to confirm it, then teaching approaches that place a heavy emphasis on the ability to blend individual sounds into wholes will fail with children experiencing speech and/or reading problems. While the child may be able to perform the task at an acoustic level, sounding out the letters c–a–t, he may be unable to synthesize the segments into the word "cat." There exists for the child no linguistic awareness of the phonemes and their relationship to the syllable and thus to the word.

A similar situation exists for the child with a speech articulation difficulty who may be quite capable of articulating the component elements of a blend s–t–r, or, b–l–u, but be unable to re-synthesize them in, for example the words "string" or "blue."

Problems of speech perception and production, and of reading may, therefore, be considered in terms of difficulty or error in language coding. If the perception of speech occurs by some active system, as illustrated by an analysis-by-synthesis model, knowledge of the appropriate generative language rules is a prerequisite to the restructuring of the message. In other words, perception entails the ability to impose syntactic structure on the acoustic or visual stimulus. To do this, the child must have acquired a correct knowledge of the rules and have gained developmental experience in their operation. Children with auditory language problems appear to have failed for one reason or another to correctly deduce the rules of the language game from the examples provided in their environment. These problems, signaling an inability to apply the rules of linguistic structure at a *specific* level, are however of an *integrative* nature.

We have seen how closely theories of reading parallel those of speech perception. In particular, we discussed the concept of reading as a deliberately acquired language-based skill dependent upon linguistic awareness. Stark (1975) has used this linguistic framework to support the claim that reading failure is therefore a language-based problem. We have already rejected the suggestion that auditory learning disability originates from difficulties with specific auditory skills. Stark likewise rejects the notion that reading problems arise primarily from difficulty with visual-perceptual skills. He points out that in a review of the studies that have evaluated the relationship between visual perception and reading ability, Hammill (1972) found results which indicated little predictive value in perceptual tests. Furthermore of 25 studies evaluated, only four provided evidence that systematic visual-motor perceptual training produced an associated improvement in reading.

Stark raises the question which we considered in the last chapter, namely, what is the relationship between the ability to perform well on tests of isolated components of perceptual processing and the ability to perform the holistic hierarchical task of interpreting spoken or written language? He reviews some of the research which has attempted to answer this question in terms of both auditory and visual-motor training (Smith and Marx, 1972; Koppitz, 1972; Burns and Watson, 1973; Hammill and Larsen, 1974) and finds the evidence essentially not supportive of an individual skills training approach to learning disabilities. Stark says:

> Some of the research we have cited suggests that children with reading failure need to learn the rules of spoken language. They need to develop strategies for processing morphophonemic and syntactic units and learn the logic of the language systems. The efficacy of training so-called "word-attack" skills, teaching sound and letter correspondences, blends, and improving perceptual motor abilities must be questioned. (1975, p. 833)

Our approach to the alleviation of these problems must therefore pay careful attention to the role that language-coding mechanisms may play in auditory learning disorders. It becomes imperative that we do everything possible to determine at what level of processing the difficulties lie.

REFERENCES

BURNS, G. W. AND B. L. WATSON, 1973. Factor analysis of the revised ITPA with underachieving children. *J. Learn. Disabil.*, **6**, 371–77.

CONRAD, R., 1972. "Speech and Reading," in *Language by Ear and by Eye*, eds. J. F. Kavanagh and I. G. Mattingly. Cambridge, Mass.: The M.I.T. Press.

CORCORAN, D. W. J., 1971. *Pattern Recognition* (Penguin Science of Behavior Series). Middlesex, England and Baltimore, Md.: Penguin Books, Ltd.

CORCORAN, D. W. J. AND R. O. ROUSE, 1970. An aspect of perceptual organization involved in the perception of handwritten and printed words. *Quart. J. Exp. Psych.*, **22**, 526–30.

EDFELDT, A. W., 1960. *Silent Speech and Silent Reading*. Chicago: Chicago University Press.

FRIES, C. C., 1962. *Linguistics and Reading*. New York: Holt, Rinehart, and Winston.

GIBSON, E., A. SHURCLIFF, AND A. YONAS, 1970. "Utilization of Spelling Patterns by Deaf and Hearing Subjects," in *Basic Studies on Reading*, eds. H. Levin and J. P. Williams. New York: Basic Books.

GOUGH, P. B., 1972. "One Second of Reading," in *Language by Ear and by Eye*, eds. J. F. Kavanagh and I. G. Mattingly. Cambridge, Mass.: The M.I.T. Press

HAMMILL, D., 1972. Training visual perceptual processes, *J. Learn. Disabil.*, **5**, 552–60.

HAMMILL, D. AND S. C. LARSEN, 1974. The effectiveness of psycho-linguistic training, *Except. Child.*, **41**, 5–16.

HARDYCK, C. D. AND L. PETRINOVITCH, 1970. Subvocal speech and comprehension level as a function of the difficulty level of reading material. *J. Verb. Learn. Verb. Behav.*, **9**, 647–52.

HORTON, K. B., 1974. "Infant Intervention and Language Learning," in *Language Perspectives—Acquisition, Retardation, and Intervention*, eds R. L. Schiefelbusch and L. L. Lloyd. Baltimore, Md.: University Park Press.

KAVANAGH, J. F. (ed.), 1968. *Communicating by Language: The Reading Process*. Bethesda, Md.: National Institute of Child Health and Human Development.

KOHLERS, P. A., 1970. "Three Stages of Reading," in *Basic Studies in Reading*, ed. J F. Kavanagh and I. G. Mattingly. New York: Basic Books.

KOPPITX, E. M., 1972. Special class pupils with learning disabilities: a five year follow up study. *Acad. Ther.*, **8**, 133–39.

LEVIN, H. B., 1972. "What the Child Knows about Speech," in *Language by Ear and by Eye*, eds. J. F. Kavanagh and I. G. Mattingly. Cambridge, Mass.: The M.I.T. Press.

LIBERMAN, A. M., 1970. The grammars of speech and language. *Cognitive Psych.*, 1, 301–23.

LIBERMAN, A. M., I. G. MATTINGLY, AND M. T. TURVEY, 1972. "Language Codes and Memory Codes," in *Coding Processes in Human Memory*, eds. A. W. Melton and E. Martin. Washington, D.C.: V. H. Winston.

LLOYD, L. L. (Ed.), 1976. *Communication Intervention and Strategies.* Baltimore, Md.: University Park Press.

MATTINGLY, I. G., 1972. "Reading, the Linguistic Process, and Linguistic Awareness," in *Language by Ear and by Eye,* eds. J. F. Kavanagh and I. G. Mattingly. Cambridge, Mass.: The M.I.T. Press.

McGUIGAN, F. J., 1970. Covert oral behavior during the silent performance of language tasks. *Psycho. Bull.,* **7,** 309–26.

MYKLEBUST, H. R., 1970. *The Psychology of Deafness.* New York: Grune and Stratton.

NOVIKOVA, L. A., 1966. "Electrophysiological Investigation of Speech," in *Thinking: Studies of Covert Language Processes,* ed. F. J. McGuigan. New York: Appleton-Century-Crofts.

REES, N. S., August 1973. Auditory processing factors in language disorders: The view from Procrustes bed. *J. Sp. Hearing Dis.,* **30,** 304–15.

SMITH, P. AND R. MARX, 1972. Some cautions on the use of the Frostig test: a factor analytic study. *J. Learn Disabil.,* **5,** 357–63.

STARK, JOEL, 1975. Reading failure: a language-based problem. *ASHA,* **17,** (12), 832–34.

STARK, R. E., 1974. *Sensory Capabilities of Hearing-Impaired Children.* Baltimore, Md.: University Park Press.

STEVENS, K. N. AND M. HALLE, 1967. "Remarks on Analysis-by-Synthesis and Distinctive Features," in *Models for the Perception of Speech and Visual Form,* ed. Wathen-Dunn. Cambridge, Mass.: The M.I.T. Press.

11

Intervention

The literature abounds with texts and articles on remedial approaches to auditory perceptual learning problems. However, these are based almost uniformly upon the assumption that the problems originate in a dysfunction of one or more of the discrete auditory abilities mentioned in Chapter 9. Only recently has the literature begun to reflect concern about the validity of hypothesizing a set of discrete auditory skills whose dysfunction results in speech, language, or reading problems.

In our selective examination of research evidence, there is a palpable absence of any well-defined theoretical model which emphasizes the relationship of discrete auditory-perceptual skills to language learning and language disorders. There exists a similar paucity of information concerning the role of auditory perception in speech articulation disorders and reading difficulties. When we examine the research findings concerning normal speech perception, we find that, rather than providing support for an auditory skills explanation, they tend to negate it.

Nevertheless, let us for a moment assume that these individual auditory skills do play a causative role in communicative disorders. We are immediately confronted with the problem of how to isolate them for independent analysis. Each time we set out to assess performance on one skill, we find that it is so interwoven with others that we cannot be sure which aspect we are testing. For example, how do we assess sequencing ability without involving the adequacy of the child's auditory attention, short-term memory, auditory discrimination, and synthesis? If we decide to test auditory memory, it is necessary for us to determine that the factors of attention, discrimination, sequencing, and synthesis are not contaminating our results.

> Because speech production is rapid, the auditory perceptual tasks of attention, focusing, tracking, discrimination, sorting, scanning, and sequencing need to be accomplished in such a brief time span that for all practical purposes they occur simultaneously. It should, therefore, be kept in mind that tasks used in studies which attempt to isolate one or another aspect of auditory perception may really be testing two or more abilities. (Witkin, 1971, p. 46)

Locke and Goldstein have emphasized that one of the major problems of assessment lies in the tendency of the tester to determine on an *a priori* basis that the particular subskill is relevant to the problem. Depressed performance on the subskill is then interpreted as being associated with the problem. While it may be related, the test data thus obtained do not justify the assumption.

We have rejected the reductionist theory that a group of independent skills underlies perception. We have instead emphasized the fact that perception is an active holistic process. The organism's perception of an object or event is heavily dependent upon how it processes the raw data. As Toch and McLean state:

> Each percept from the simple to the most complex is the product of a creative act. . . . The raw material for creation is lost to us, since in the very act of creating we modify it. (1970, p. 127)

Piaget has stressed the same active role:

> . . . In the end, the relative adequacy of any perception to any object depends on a constructive process and not on an immediate contact. During this constructive process the subject tries to make use of whatever information he has, incomplete, deformed or false as it may be, and to build it into a system which corresponds as nearly as possible to the properties of the object. (1969, p. 365)

Perception as a holistic concept forces us away from the assessment and remediation of individual skills and towards an integrative systems approach. Our concern is for *how a child is performing a task*. This is in contrast to asking how deviant his behaviour is on a test assumed to assess a component we judge to be critical to adequate performance. Muma and Lubinski (unpublished manuscript) apply this concept to intervention procedures. They emphasize the need to identify relevant intervention alternatives rather than to resolve aberrance in normative assessment. They distinguish between the collection of normative data that compares the individual to the group, and the collection of evidence.

> Data become evidence only when they are consistent (by verification or prediction) with theoretical premises and constructs. What the teacher/clinician needs to know about the child's aberrant performance is not how aberrant it is, but what is the nature of the processes which underlie the speech or reading behavior. The concern must be for how the *child* generates his spoken language patterns.

If we accept the generative model, we must acknowledge that perception and production of spoken and written language is based upon a familiarity with the rules of the language culture: you cannot do it until you know how. Conversely, the ability to process language correctly is evidence of that knowledge. Continuing this line of reasoning, inability to process language appropriately must indicate either that the child has not yet acquired a knowledge of the rules of the system, or that at some level he has acquired a different set of conditions for their application.

Muma and Lubinski state:

> The specification of linguistic and referential conditions has direct clinical implications because it is directly relevant to how a particular individual uses or functions with a particular verbal system or process. Such *evidence* offers opportunities to deduce appropriate intervention alternatives.

Weener has hypothesized that, while children with normal language processing systems are able to actively apply linguistic strategies to the integration of the incoming auditory information, the child with a learning disability has only a limited number of such processing patterns in his repertoire. In discussing his approach to remediation, he states:

> One of the basic principles underlying my proposals for remediation is that the primary difference between the person with good auditory comprehension and the person with poor auditory comprehension is in the nature and use of the strategies, plans, hypotheses, and rules that are available to the person. Remediation programs would then

teach children to use a structured pattern which they already have, or to learn new patterns which can serve as the basis for structuring new material. (1972, p. 164)

When we consider intervention approaches we cannot ignore the research evidence and theoretical models of speech processing we have discussed. Most important among the findings is the hierarchical relationship which links together all levels of the processing of oral/auditory language. If we accept this hierarchical model, it is logical to assume that a deviancy of performance at one level will reflect on and be evidenced in, performance at other levels. Instead of evaluating and working only at the level of observed deviance, it becomes important to carry out a more comprehensive assessment of the efficiency of the entire system. Even within a given level, for example the phonetic level, the pattern of production of a given phoneme is known to vary with phonetic context. We have discussed the fact that the minimal unit of speech production and perception is almost always at least a syllable and usually is even longer than a syllable. Thus, the pattern of production for any phoneme must change to accommodate the transitional characteristics of the context in which it occurs.

As Martin (1974) has pointed out, a task which assesses phonemic discrimination in a nonchanging context does not provide an evaluation of auditory perceptual skills involved in speech perception because the auditory system must be capable of dealing with the acoustic variations arising from contextual influences. Similarly, you will recall that Blesser's (1974) results indicate that phoneme perception is not a prerequisite to word recognition since if the word pattern can be identified, processing can occur at a higher level than the phonological level.

The research of Menyuk (1964); Shriner, Holloway, and Daniloff (1969); Locke (1970); and Locke and Goldstein (1970), discussed in Chapter 9 supports the assumption that language encoding/decoding dysfunction often underlies the more peripheral phonetic encoding/decoding difficulties manifest by children with multiple errors of speech articulation. These findings are not surprising if we accept the suggestion that speech and language are processed according to a common grammar. As the results of the studies suggest, such a premise would lead to the conclusion that many speech articulation problems have their roots on the level of linguistic rather than acoustic processing. The differentiation of these two levels of processing is critical to intervention procedures, as Ingram points out:

Linguistic perception adds the requirements of categorizing the signal perceptually into representations of meaningful elements. This abstraction process results in reducing radically the otherwise effective ability of acoustic perception. (1974, p. 331)

When we seek to identify a deviant pattern of spoken language processing in a child, we cannot assess only primary acoustic processing skills, nor can we ignore differences between primary and linguistic processing. Testing and evaluation should be directed at determining *phonemic* and *syntactic* discrimination as well as *phonetic* discrimination and should do so over units syllable size or greater. The effects of *contextual* influence also need to be determined.

If, in fact, the nature of the difficulty is rule-based, then only by determining the rule governing the child's response behavior can we begin to understand the implications of symptoms arising from the operation of that rule. Thus, any attempt to assess the child's perception of speech sounds should be in terms of his own rule system over a range of contexts.

Whitacre, Luper, and Pollio (1970) have also determined that children with defective speech articulation evidenced general language deficits. They argued that recognition of this fact in intervention approaches should be directed at both the linguistic and the phonetic processing of such children. In the words of Marquardt and Saxman, teachers and therapists should recognize that:

. . . if the child is deficient in the area of language, the therapy program should not be limited to speech sound acquisition. (1972, p. 387)

If, in fact, the errors of phonemic production are related to the processing of language at higher levels, as Shriner et al. (1969) have suggested, enrichment of the child's syntactic processing should increase responsiveness to articulation therapy. It makes little sense to pursue a course of intensive auditory discrimination with a speech-defective child whose problem lies above the primary acoustic level. If it can be demonstrated that a child not only manifests deviant speech articulation but also depressed syntactic processing, remedial training should be emphasized at the level of syntax.

In a similar vein, when auditory memory appears to be deficient, remedial efforts should be directed at teaching transformational structures and grammatical rules. The assumption being that this will facilitate the retention of larger linguistic units and strengthen long-term memory of the correct rules for generating appropriate structures.

Thus, unless proven otherwise, we should assume that nonorganic, multiple disorders of speech articulation are also deficient in higher levels of language processing. Therefore, syntactic competency and proficiency should be evaluated (Marquardt and Saxman, 1972).

We have been discussing approaches to the management of children who evidence disorders of speech articulation. It has been suggested that frequently these disorders have their origins in language processing difficulties rather than in deficits of specific auditory skills. Similarly, it has been argued that reading is also a language-based process. It is not surprising, therefore, that poor reading abilities are not uncommon among children with multiple speech articulation problems. This in turn results in difficulties in learning through the medium of the printed word. Nevertheless, it cannot be assumed that a speech difficulty is synonymous with learning disability. It is essential to differentiate those children whose speech articulation errors are in fact symptomatic of problems in the higher levels of language processing with resultant impairment of learning ability.

Vogel (1974) and Semel and Wiig (1975) have demonstrated that children with learning difficulties exhibit a reduction in the comprehension and use of syntactic forms which is also reflected in reading abilities. Thus, while it is recognized that the majority of children who experience learning difficulties do not manifest speech problems, it is suggested that a language-based approach to their problems is equally appropriate.

In discussing the results of their study of children with learning disabilities, McGrady and Olsen (1970) argue that a remedial program based upon perceptual training alone is unwarranted. Their test data indicate that children who fail in school because of verbal difficulties (usually indicated by failure to learn to read, or failure to read to grade level) primarily experience language processing difficulties. The authors therefore urge that intervention be approached from a total language standpoint. They stress that, since verbal problems are not confined to any specific psychosensory modality, perceptual training aimed at improving specific modal skills is unlikely to be markedly successful.

Current research indicates that the type of problems we have been discussing are symptomatic of deviancies in underlying generative language processes. We are also assuming that verbal behavior is rule-based from the phonemic to the cognitive level. If this is the case, then we must conclude that a deficiency of language patterning arises either from failure to acquire a particular rule or from faulty rule acquisition. Intervention procedures must therefore aim at determining the nature of the underlying rule governing the deviant pattern. Intervention must seek effective modification to bring about, at all levels, conformity with standard language structure. The procedure involves

1. Careful description of the child's language processing behavior
2. Identification of patterns derived from observation across phonetic, syntactic, and semantic contexts
3. Formulation of an hypothesis as to which rule(s) is operant in the child's recognition and/or production of the deviant pattern
4. Identification of an appropriate intervention strategy designed to effect a modification or substitution for the incorrect rule.

The information must be assembled under meaningful contingencies that yield insight into how and why a child is failing to correctly perceive and/or generate a particular linguistic pattern. This is important, for as Muma and Lubinski (unpublished manuscript) explain,

> The specification of linguistic and referential conditions has direct clinical implications because it is directly relevant to how a particular individual uses or functions with a particular verbal system or process. Such *evidence* offers opportunities to deduce appropriate intervention alternatives.

The importance of sampling the child's speech is illustrated by evidence that we cannot predict an underlying rule from a single occurrence of an unacceptable grammatical structure. This holds true for all levels (McDonald, 1964b; Bloom, 1974). Bloom points out that

> . . . both understanding and use of a particular linguistic form—whether a particular word, a grammatical morpheme or a syntactic structure—depend on a great many variables, both linguistic and nonlinguistic. (1974, p. 297)

The absence of a particular form in a speech sample does not necessarily mean that the rule for its generation is totally absent from the child's language system. Nor, conversely, can evidence of a child's use of a particular form in an isolated speech sample be used to support the assumption that the form is understood and used in a wide variety of contextual situations. Bloom explains that a child's speech is modified by its context. She presents evidence to show that there is often a heavy dependence upon cognitive appreciation of the situation in which the form occurs and the state of affairs to which it refers. Her subject, a child aged 32 months, 2 weeks demonstrated that:

When asked to reproduce sentences that did not relate to immediate context and behavior, he could not produce the very same uttterance he himself had produced with such support. (1974, p. 300)

This example emphasizes the relationship which exists between cognition and spoken language processing. In both assessment and intervention, it is of great importance, therefore, to work with stimuli that are clear, meaningful, and immediately relevant to the child. The unit size should be sufficient to allow for the operation of maximal co-articulatory function and contextual influences to facilitate recognition and production strategies (Shriner et al. 1969, p. 324).

It is beyond the scope of this text to outline specific strategies for the development of intervention programs with children exhibiting various auditory learning difficulties. The literature in this area is limited, for the approach is still in an inceptual stage. Specific applications of the theoretical constructs to language processing difficulties are still being explored. The implications of these theories for intervention strategies for helping children with learning difficulties lag even further behind theoretical models. We must await further research into remedial procedures and the collection, documentation, and evaluation of these data before generalized therapeutic methods can be developed.

Ayers (1975) has discussed the limitations of studies which have sought to demonstrate the effectiveness of perceptual motor training in helping such children. She lists three major factors: a) the tendency to list a few isolated procedures derived from a larger theoretical framework; b) insufficient understanding of the theory being used; c) a failure to match the intervention program to the child's particular type of difficulty. Ayers deplores the tendency to consider all learning disorders to be alike and the subsequent use of a standard intervention approach. She argues that successful intervention programs were generally found to be those which were soundly based on a well-developed theoretical rationale implemented by teachers thoroughly familiar with the theory. The studies which were less successful were those which adopted isolated procedures as a "recipe" rather than using a dynamic monitored approach based on knowledge of sensori-motor development, the nature of the problem, and the child's response.

Much information is available concerning the nature and operation of strategies underlying the processing of verbal behavior. Schiefelbusch and Lloyd (1974) in their edited text *Language Perspectives—Acquisition, Retardation, and Intervention* and Lloyd's edited text *Communication Assessment and Intervention (1976)* have performed a valuable contribution by pulling this information together. The two-volume work *Perceptual and Learning Disabilities in Children* (Cruickshank and Hallahan, 1975) affords a rich source of information concerning the broader aspects of learning disabilities. They examine current psychoeducational practices (Vol. 1) and research and theory (Vol. 2). Further texts and journal articles may be expected to delve deeper into the methodology and remediation of the more subtle problems which, it has been suggested, underlie these learning disabilities.

It has been our limited purpose in the latter section of this text to explore the possible implications which a pattern-processing model of speech perception might have for the management of verbal communication disorders in children. The model seems to accommodate the research evidence in this area surprisingly comfortably. While this does not confirm the validity of the model, it does suggest that this approach will provide us with a central focus necessary to unify the many facets of the problem. It appears that a pattern-processing model affords a way of looking at auditory processing of speech which can accommodate both normal and deviant function. If this is so, perseverance with the building of our case is more than justified.

For the teacher or clinician, the value of any model of spoken language processing rests in its usefulness as a tool. Regardless of the conceptual integrity and appeal of the theory represented by the model, it must prove effective in furthering the search for answers to the speech, language, and learning problems experienced by so many children. The value of the model constructed in this text can be judged only in terms of how much it contributes to *your* insight into the problems *you* seek to remediate.

REFERENCES

Ayers, A. J., 1975. "Sensorimotor Foundations of Academic Ability," in *Perceptual and Learning Disabilities in Children*, eds. W. M. Cruickshank and D. P. Hallahan, **2**. Syracuse, N.Y.: Syracuse Univ. Press.

Bloom, L., 1974. "Talking, Understanding, and Thinking," in *Language Perspectives—Acquisition, Retardation, and Intervention*, eds. R. L. Schiefelbusch and L. L. Lloyd. Baltimore, Md.: University Park Press.

Cruickshank, W. M. and D. P. Hallahan, eds., 1975. *Perceptual and Learning Disabilities in Children: 1 Psycho-educational Practices, 2 Research and Theory*. Syracuse, N.Y.: Syracuse Univ. Press.

INGRAM, D., 1974. "The Relationship between Comprehension and Production," in *Language Perspectives—Acquisition, Retardation, and Intervention*, eds. R. L. Schiefelbusch and L. L. Lloyd. Baltimore, Md.: University Park Press.

LOCKE, J. L., 1971. The child's acquisition of phonetic behaviour. *Acta Symbolica*, **2**, (1), 28–32.

LOCKE, J. L. AND J. I. GOLDSTEIN, 1971. Children's identification and discrimination of phonemes. *British Journal of Communication Disorders*, **6**, 107–12.

LOCKE, J. L. AND J. I. GOLDSTEIN, 1973. Children's attention and articulation. *Lang. Speech*, **16**, (2), 156–68.

MARQUARDT, T. P. AND J. H. SAXMAN, 1972. Language comprehension and auditory discrimination in articulation deficient kindergarten children. *J. Sp. Hrg. Res.*, **15**, (2), 383–89.

MARTIN, A. D., 1974. Some objections to the term "Apraxia of Speech." *J. Sp. Hrg. Dis.*, **39**, (1), 53–64.

McDONALD, E. T., 1964. *Articulation Testing: A Sensory Motor Approach*. Pittsburgh, Pa.: Stanwix House.

McDONALD, E. T., 1964b. *A Deep Test of Articulation*. Pittsburgh, Pa.: Stanwix House.

McGRADY, H. J. AND D. A. OLSEN, April, 1970. Visual and auditory learning processes in normal children and children with specific learning disabilities. *Exceptional Children*, 581–88.

MENYUK, P., 1964. Comparison of grammar of children with functional deviant and normal speech. *J. Sp. Hrg. Res.*, **7**, 109–21.

MUMA, J. AND R. LUBINSKI (unpublished manuscript). "Data or Evidence."

PIAGET, J., 1969. *The Mechanisms of Perception*. New York: Basic Books.

REES, N. S., August, 1973. Auditory processing factors in language disorders: The view from Procrustes bed, *J. Sp. Hrg. Dis.*, **38** (3), 304–15.

SCHIEFELBUSCH, R. L. AND L. L. LLOYD (eds.), 1974. *Language Perspectives-Acquisition, Retardation, and Intervention*. Baltimore, Md.: University Park Press.

SHRINER, T. H., M. S. HOLLOWAY, AND R. G. DANILOFF, 1969. The relationship between articulatory defects and syntax in speech defective children. *J. Speech and Hearing Res.*, **12**, (2), 319–25.

TOCH, H. AND M. S. McLEAN, 1970. "Perception and Communication: A Transactional View," in *Foundations of Communication Theory*, ed. K. K. Sereno and C. D. Mortensen. New York: Harper & Row.

WEENER, P., 1972. "Toward a Developmental Model of Auditory Processes," in *The Proceedings of the First Annual Memphis State University Symposium on Processing and Learning Disabilities*, ed. D. L. Rampp. Memphis, Tenn.: Memphis State Univ. Press.

WHITACRE, J. D., H. L. LUPER, AND H. R. POLLIO, 1970. General Language deficits in children with articulation problems. *Lang. Speech*, **13**, 231–39.

WITKIN, B. R., 1971. "Auditory Perception—Implications for Language Development," *Language Speech and Hearing Service in Schools*, **4**, 31–52. Washington, D.C.: American Speech and Hearing Assoc.

AUTHOR INDEX

241

SUBJECT INDEX